AF388985

Neuromodulation for Intractable Pain

Neuromodulation for Intractable Pain

Special Issue Editors

Tipu Aziz
Alex Green

MDPI • Basel • Beijing • Wuhan • Barcelona • Belgrade • Manchester • Tokyo • Cluj • Tianjin

Special Issue Editors

Tipu Aziz
University of Oxford
UK

Alex Green
University of Oxford
UK

Editorial Office
MDPI
St. Alban-Anlage 66
4052 Basel, Switzerland

This is a reprint of articles from the Special Issue published online in the open access journal *Brain Sciences* (ISSN 2076-3425) (available at: https://www.mdpi.com/journal/brainsci/special_issues/neuromodulation_pain).

For citation purposes, cite each article independently as indicated on the article page online and as indicated below:

LastName, A.A.; LastName, B.B.; LastName, C.C. Article Title. *Journal Name* **Year**, *Article Number*, Page Range.

ISBN 978-3-03936-950-8 (Hbk)
ISBN 978-3-03936-951-5 (PDF)

Contents

About the Special Issue Editors

Tipu Aziz, D Med Sci, FRCS(SN), is the founder and head of Oxford functional neurosurgery. His primate work was central to confirming the subthalamic nucleus as a possible surgical target for deep brain stimulation in Parkinson's disease and more recently the pedunculopontine nucleus. OFN is currently one of the busiest centres for such surgery in the UK and academically very productive. Research Interests are the role of the upper brain stem in the control of movement, the clinical neurophysiology of movement disorders and neuropathic pain and autonomic responses to deep brain stimulation, use of MR and MEG imaging in functional neurosurgery.

Alex Green MD FRCS(SN)., has been looking at the neurocircuitry underlying autonomic function and pain in humans undergoing Deep Brain Stimulation (DBS) over the past ten years. There are several aims of this research. Firstly, he wishes to understand both the mechanisms underlying the pathophysiology of neuropathic pain as well as why some patients get much better than results than others. Secondly, by understanding the autonomic nervous system, it may be possible to control diseases such as hypertension, respiratory and bladder disease by brain manipulation in the future. Most of the research to date has involved stimulating brain areas under different experimental conditions and also recording local field potentials to understand the underlying neurophysiology. This work has resulted in a number of publications including improvement in peak expiratory flow with stimulation, the effect of stimulation on blood pressure and baroreceptors sensitivity and novel electrical signals associated with pain states.

Editorial

Neuromodulation for Intractable Pain

Alexander L. Green * and Tipu Z. Aziz

Nuffield Department of Surgical Sciences, University of Oxford, Oxford OX3 9DU, UK; tipu.aziz@nds.ox.ac.uk
* Correspondence: alex.green@nds.ox.ac.uk

Received: 22 April 2020; Accepted: 29 April 2020; Published: 30 April 2020

Over 7% of the Western population suffer from intractable pain and despite pharmacotherapy, many patients' pain is refractory [1]. In addition to the pain, patients often suffer from depression and anxiety, poor quality of life and loss of employment. An ever-enlarging problem is that of opiate use, which in the US has been labelled as a "crisis" [2]. In order to tackle these issues, we require a greater understanding of the underlying pathophysiology of pain, novel treatments (pharmacological and otherwise), and a greater evidence base for both the efficacy of non-pharmacological treatments alongside a better understanding of the mechanisms of action. In this issue, Deer et al. [3] provide an up-to-date literature review on spinal cord stimulation (SCS), dorsal root ganglion (DRG) stimulation, and peripheral nerve stimulation (PNS), which are all well-established neuromodulatory techniques for treating chronic neuropathic pain. Deer et al. provide a comprehensive report, demonstrating that SCS has well-established efficacy for specific pain subtypes such as failed back surgery syndrome (FBSS), complex regional pain syndrome (CRPS), and a number of other conditions. They point out that although SCS is not a new therapy, there are a multitude of new advancements in the field such as novel waveforms, new closed-loop technologies, and many recent advances in the understanding of its mechanisms. Whilst DRG stimulation and PNS are somewhat more recent additions to the armamentarium, there is good early evidence for efficacy, although the authors point out that trial designs (especially subject blinding) can be a challenge. Dones and Levi, in their review of SCS, echo the conclusions of Deer et al. and also discuss in depth the technical nuances of SCS therapy. Controversies include the choice between percutaneous and paddle electrodes, and the choice between awake implantation and implantation under general anaesthetic. The authors present the evidence on different sides of the argument, providing the advantages and disadvantages of each technique. This also makes the point that trials need to be evaluated in the context of the specific technique. Regarding the mechanisms of action of DRG stimulation, Parker et al. [4] report a study in which magnetoencephalography (MEG) was used to measure cortical activity during periods of DRG stimulation compared with a control whilst performing a cognitive task (the "N-Back task"). The authors elegantly show that DRG stimulation modulates cortical gamma activity in the cognitive dimension of pain. This study has implications for the way in which peripheral neuromodulation works and implies that the modulation of cortical networks is important (either as a cause or consequence), and not just local DRG effects. Salgado et al. [5], in their study on CRPS in mice, bring to our attention that there are alternatives to medication, other than neuromodulation. One such intervention is manual therapy such as ankle joint mobilization. The authors show that mobilization 48 hours after an ischemia–reperfusion injury reduced the pain behaviour and oxidative stress. This study outlines the importance of therapy in the acute phase after injury in order to prevent the build-up of chronic pain in the first place.

For those patients who do not respond to SCS and other forms of more "peripheral" neuromodulation, deep brain stimulation (DBS) and motor cortex stimulation (MCS) are alternatives. Farrell and colleagues [6] review the history and literature on these treatments and conclude that whilst there are many studies showing efficacy, there is a lack of well-designed clinical trials and that more work is needed to assess the factors that predict success in individual patients. Farrell et al. also summarise a newer target for DBS for pain: the anterior cingulate cortex (ACC). Further work on ACC DBS for chronic pain is highlighted by Huang et al [7]. Their study follows an individual who gained successful pain relief with bilateral ACC DBS but unfortunately also developed disabling generalised seizures that were related to the stimulation amplitude. By applying a novel brain recording device (Medtronic PC + S®, Minneapolis, MN, USA), the authors were able to identify the patterns of stimulation that precluded the seizure activity. This is a prime example of how evolution in device technology can enable successful treatment in patients that have been deemed "untreatable" with existing technology. In a second study, Farrell et al. [8] highlight the use of DBS for a range of pain and non-pain conditions. The latter concentrates mainly on autonomic symptoms such as hypertension and bladder symptoms, often investigated in the context of DBS for existing conditions such as Parkinson's disease. The authors point out that DBS is a useful treatment for a range of chronic symptoms that cause suffering and that the realm of palliative care is not just for patients with a limited life expectancy.

In addition to studies looking at neuromodulation as a general treatment for refractory conditions, more work is needed into its use in specific pain syndromes. Roy et al. [9] summarise pelvic and urogenital pain and the use of neuromodulation in its management. The authors demonstrate that the neurocircuitry underpinning the pelvic and urogenital system may be targeted from peripheral (e.g., posterior tibial or pudendal nerves) to central (periaqueductal grey area). Again, there are many gaps in our knowledge regarding both mechanisms of action and efficacy. There is also much more work needed to understand the underlying molecular changes in pain sub-types that will help inform drug design but also influence the targets for neuromodulation. Lombardo et al. present an intriguing study looking at the interleukin-1 receptor antagonist (IL-1RN) expression in a murine cortical spreading depression (CSD) model of migraine [10]. The investigators demonstrate that there is an upregulation of IL-1RN and hypothesise that this demonstrates a possible attempt to modulate the inflammatory response. The link between chronic pain and the immune system is gaining increasing interest in the literature and it is likely that further investigation is important for both chronic pain management and the tentative possibility of using neuromodulation to alter the immune response, as is already being investigated in relation to vagal nerve stimulation [11].

Conflicts of Interest: The authors declare no conflict of interest.

References

1. Torrance, N.; Smith, B.H.; Bennett, M.I.; Lee, A.J. The Epidemiology of Chronic Pain of Predominantly Neuropathic Origin. Results From a General Population Survey. *J. Pain* **2006**, *7*, 281–289. [CrossRef] [PubMed]
2. Ostling, P.S.; Davidson, K.S.; Anyama, B.O.; Helander, E.M.; Wyche, M.Q.; Kaye, A.D. America's Opioid Epidemic: A Comprehensive Review and Look into the Rising Crisis. *Curr. Pain Headache Rep.* **2018**, *22*, 32. [CrossRef] [PubMed]
3. Deer, T.; Jain, S.; Hunter, C.; Chakravarthy, K.V. Neurostimulation for Intractable Chronic Pain. *Brain Sci.* **2019**, *9*, 23. [CrossRef] [PubMed]
4. Parker, T.; Huang, Y.; Raghu, A.L.; Fitzgerald, J.J.; Green, A.L.; Aziz, T.Z. Dorsal Root Ganglion Stimulation Modulates Cortical Gamma Activity in the Cognitive Dimension of Chronic Pain. *Brain Sci.* **2020**, *10*, 95. [CrossRef] [PubMed]
5. Salgado, A.S.I.; Stramosk, J.; Ludtke, D.D.; Kuci, A.C.C.; Salm, D.C.; Ceci, L.A.; Petronilho, F.; Florentino, D.; Danielski, L.G.; Gassenferth, A.; et al. Manual Therapy Reduces Pain Behavior and Oxidative Stress in a Murine Model of Complex Regional Pain Syndrome Type I. *Brain Sci.* **2019**, *9*, 197. [CrossRef] [PubMed]
6. Farrell, S.M.; Green, A.; Aziz, T. The Current State of Deep Brain Stimulation for Chronic Pain and Its Context in Other Forms of Neuromodulation. *Brain Sci.* **2018**, *8*, 158. [CrossRef] [PubMed]
7. Huang, Y.; Cheeran, B.; Green, A.L.; Denison, T.J.; Aziz, T.Z. Applying a Sensing-Enabled System for Ensuring Safe Anterior Cingulate Deep Brain Stimulation for Pain. *Brain Sci.* **2019**, *9*, 150. [CrossRef] [PubMed]
8. Farrell, S.M.; Green, A.L.; Aziz, T.Z. The Use of Neuromodulation for Symptom Management. *Brain Sci.* **2019**, *9*, 232. [CrossRef] [PubMed]

9.   Roy, H.; Offiah, I.; Dua, A. Neuromodulation for Pelvic and Urogenital Pain. *Brain Sci.* **2018**, *8*, 180. [CrossRef]
      [PubMed]
10.   Lombardo, S.D.; Mazzon, E.; Basile, M.; Cavalli, E.; Bramanti, P.; Nania, R.; Fagone, P.; Nicoletti, F.; Petralia, M.
      Upregulation of IL-1 Receptor Antagonist in a Mouse Model of Migraine. *Brain Sci.* **2019**, *9*, 172. [CrossRef]
      [PubMed]
11.   Huffman, W.J.; Subramaniyan, S.; Rodriguiz, R.M.; Wetsel, W.; Grill, W.M.; Terrando, N. Modulation of
      neuroinflammation and memory dysfunction using percutaneous vagus nerve stimulation in mice. *Brain
      Stimul.* **2019**, *12*, 19–29. [CrossRef] [PubMed]

Article

Dorsal Root Ganglion Stimulation Modulates Cortical Gamma Activity in the Cognitive Dimension of Chronic Pain

Tariq Parker *, Yongzhi Huang, Ashley L.B. Raghu, James J. FitzGerald, Alexander L. Green and Tipu Z. Aziz

Nuffield Department of Surgical Sciences, University of Oxford, Oxford OX3 9DU, UK;
yongzhi.huang@nds.ox.ac.uk (Y.H.); ashley.raghu@nds.ox.ac.uk (A.L.B.R.); james.fitzgerald@nds.ox.ac.uk (J.J.F.);
alex.green@nds.ox.ac.uk (A.L.G.); tipu.aziz@nds.ox.ac.uk (T.Z.A.)
* Correspondence: tariq.parker@linacre.ox.ac.uk or tq.parker18@gmail.com; Tel.: +44-0-7752353043

Received: 12 January 2020; Accepted: 10 February 2020; Published: 11 February 2020

Abstract: A cognitive task, the n-back task, was used to interrogate the cognitive dimension of pain in patients with implanted dorsal root ganglion stimulators (DRGS). Magnetoencephalography (MEG) signals from thirteen patients with implanted DRGS were recorded at rest and while performing the n-back task at three increasing working memory loads with DRGS-OFF and the task repeated with DRGS-ON. MEG recordings were pre-processed, then power spectral analysis and source localization were conducted. DRGS resulted in a significant reduction in reported pain scores (mean 23%, $p = 0.001$) and gamma oscillatory activity ($p = 0.036$) during task performance. DRGS-induced pain relief also resulted in a significantly reduced reaction time during high working memory load ($p = 0.011$). A significant increase in average gamma power was observed during task performance compared to the resting state. However, patients who reported exacerbations of pain demonstrated a significantly elevated gamma power ($F(3,80) = 65.011612$, $p < 0.001$, adjusted p-value = 0.01), compared to those who reported pain relief during the task. Our findings demonstrate that gamma oscillatory activity is differentially modulated by cognitive load in the presence of pain, and this activity is predominantly localized to the prefrontal and anterior cingulate cortices in a chronic pain cohort.

Keywords: Pain; Dorsal root ganglion stimulation; cognition; gamma; MEG

1. Introduction

Pain is a multi-dimensional experience, traditionally described as consisting of sensory, affective and cognitive domains [1]. Each domain can contribute to the modulation, and at times the propagation, of chronic pain. The cognitive dimension of pain has been demonstrated by investigating the roles that attention, distraction and memory play in altering pain perception [2,3]. Studies have shown that engaging attentional networks with cognitive loads can attenuate perceived pain for a given stimulus — a distraction mechanism of pain relief [4,5]. Conversely, it has also been demonstrated that pain can have a detrimental effect on attentional task performance — a disruptive effect of pain on cognition [6,7]; suggestive of an integrated network involving prefrontal, somatosensory and limbic cortices, and a complex interplay between pain and cognition among these regions.

The role of neurophysiology in these processes has revealed a similarly overlapping feature of pain and cognition—cortical gamma oscillations. High-frequency gamma activity has long been associated with cognition and attention [8,9] but has also been shown to encode ongoing pain [10,11]. Moreover, surgically implanted devices such as spinal cord stimulation have shown the potential to modulate cortical gamma (30–45 Hz) activity [12], supporting the hypothesis of supraspinal mechanisms of action for spinal, and potentially peripheral, neuromodulation.

A key structure of the peripheral nervous system, the dorsal root ganglion (DRG), contains a collection of primary afferent cell bodies in the lateral epidural space which synapse within the spinal cord laminae to convey nociceptive inputs which form the ascending spinothalamic tract. Dorsal root ganglion stimulation (DRGS) is a technique that has gained popularity over the past decade as an effective target of neuromodulation in chronic neuropathic pain and has demonstrated the potential to improve the cognitive-affective dimensions of pain [13]. Neuroimaging has been an invaluable tool to corroborate the effects of cognitive modulation in pain research [14–17]. As such, we have employed the technique of magnetoencephalography (MEG), coupled with a well-validated working memory task, the *n-back task* [18,19], to investigate the effect of DRGS-mediated pain relief on cognitive performance, the effect of increasing attentional load on the pain percept and the neurophysiologic representation of gamma-band oscillations in a cohort of chronic pain patients.

2. Materials and Methods

2.1. Participants

The study was conducted with approval from the South-Central Oxford Research Ethics Committee (REF. 13SC0298) in accordance with the Declaration of Helsinki. Sixteen patients were recruited for the study who had undergone surgical implantation of DRG stimulators at the John Radcliffe Hospital for medically refractory chronic pain syndromes (see Table 1). Informed consent was obtained, and participants were randomized, by flipping a coin, to begin MEG recordings in the *ON*-stimulation or *OFF*-stimulation condition, to counter order effects.

Table 1. Patient demographics and DRG stimulation parameters, CRPS—Complex regional pain syndrome.

Patient	Age	Gender	Diagnosis	Electrode Location	Stimulation Parameters (Frequency (Hz)/Amplitude (mA)/Pulse Width (μs))
1	49	Female	Postherpetic neuralgia	Right L5	20/1.6/400
2	53	Female	Meralgia paresthetica	Right L2	20/0.6/300
3	29	Male	Post-traumatic compressive neuropathy	Left L2	20/0.7/250
4	78	Male	Diabetic neuropathy	Bilateral L5	Right - 20/1.025/450 Left - 20/0.775/480
5	46	Male	CRPS	Right L3	20/0.7/410
6	52	Male	Post-operative nerve entrapment	Left L1	28/1.3/250
7	58	Female	CRPS	Right L2/L3	20/2.1/250
8	61	Male	Post-operative mononeuropathy	Left L3	20/2.1/140
9	47	Male	CRPS	Left L4	20/6/350
10	55	Male	Nerve entrapment	Right C7/C8	20/0.425/300
11	29	Male	Post-operative radiculopathy	Bilateral L5	Right - 20/2.25/700, Left - 20/650/800
12	52	Female	CRPS	Right L5	30/0.7/500
13	77	Female	Postherpetic neuralgia	Right T1	30/0.4/300
14	22	Female	Dystonic pain	Right L2/L3	20/2.4/300
15	52	Male	Post-operative mononeuropathy	Right L1	30/0.525/400
16	54	Male	Post-operative radiculopathy	Right L3/L4	20/0.475/360

2.2. Surgical Procedure

The DRG stimulators were implanted under local anaesthetic with light sedation (propofol) in the prone position. Under fluoroscopic control, a delivery sheath was used to enter the epidural space, and a DRG Axium® lead (Abbott Laboratories, Sunnyvale, CA, USA) was introduced under X-ray guidance into the appropriate nerve root exit foramen, so that the electrode contacts were positioned over the dorsum of the DRG in the dorsal part of the foramen. Sedation was weaned and the leads were tested for efficacy prior to re-sedation. Subsequently, when anteroposterior and lateral X-rays had confirmed satisfactory position (See Figure 1), a strain-relief loop was fashioned in the spinal canal, and the wires were tunnelled to an implantable pulse generator (IPG) that was placed subcutaneously remote from the spine.

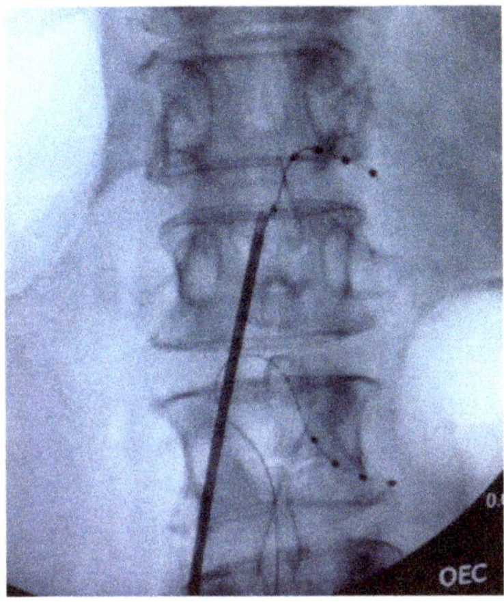

Figure 1. Fluoroscopic image of intra-operative dorsal root ganglion (DRG) lead placement at T12 and L2 on the right side.

2.3. Attentional Task

A numerical n-back task was used, which consisted of integers ranging from one to four, flashing on a display for 500 msec. Participants were instructed that three working memory loads of increasing difficulty would be cycled for the duration of the task: 0-back, 1-back and 2-back conditions. During the 0-back (low working memory) condition, participants were to immediately respond with a button press corresponding to the number flashed on screen. During the 1-back condition (low-to-intermediate working memory), participants were only to button press if the number flashing on screen corresponded to the number that flashed previously (one back). In the 2-back condition (high working memory), participants were only to button press if the number that flashed on-screen corresponded to the number that appeared two sequences before (two back).

Six trials of each condition would cycle sequentially for a total duration of twelve minutes while MEG signals were recorded. Participants were trained until they were comfortable with the paradigm and randomized to start the task in the ON or OFF stimulation condition. The possible outcomes of the task would be a "hit" (correctly identifying a target for the relevant task condition), an error of omission (failure to identify a target for the relevant task condition), an error of commission (incorrectly identifying a non-target as a target in the relevant condition) or no button press (correctly omitting a non-target) (See Figure 2).

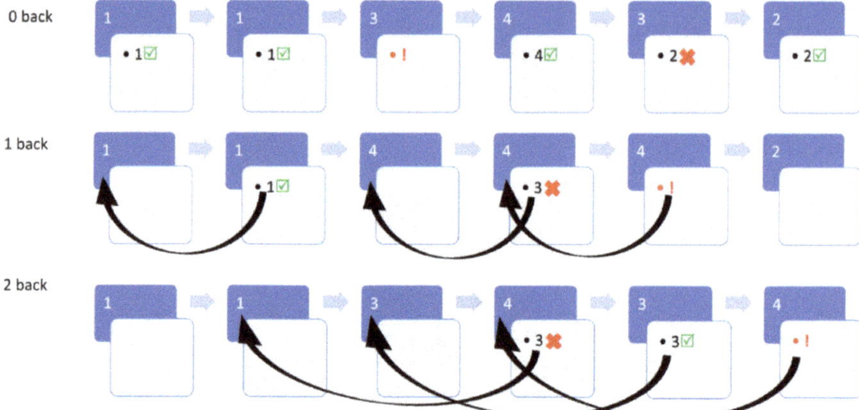

Figure 2. Diagrammatic illustration of numerical n-back task, depicting hits (☑), errors of omission (!) and errors of commission(✗) at three working memory loads (0-back, 1-back and 2-back).

Average reaction time (RT) and accuracy (number of hits/total number of targets) for each condition were calculated and evaluated for statistical differences.

2.4. Magnetoencephalography

Recordings were performed at the Oxford Centre for Human Brain Activity (OHBA) using a 306-channel Elekta Neuromag MEG system comprised of 102 magnetometers and 204 planar gradiometers at a sampling rate of 1000 Hz. The patient was relaxed and seated under the device, and the relative head position was determined and tracked using Standard Elekta-Neuromag head position indicator (HPI) during the scan. Prior to data acquisition, the HPI coil locations, the position of three anatomical landmarks (the nasion, and left and right pre-auricular points), and the head shape were measured using a three-dimensional digitizer (Polhemus Isotrack). Patients were scanned during the n-back task for 12 min in both DRGS-ON and DRGS-OFF conditions, separated by a pre-defined washout period [20] to prevent carryover effects. Patients were also scanned with the DRGS-OFF at rest with eyes open for comparison with task conditions. Electrocardiographic (ECG) recordings were monitored by applying bilateral electrodes to the volar aspect of the wrists and, simultaneously, electrooculographic (EOG) traces were recorded by two electrodes, placed above and below the left eye.

2.5. Spectral and Source Analysis

Data were visually inspected and artefacts such as flats and jumps were detected in each channel and marked. The strong magnetic artefacts in the raw data, such as the artefacts of stimulation, were suppressed by the spatiotemporal signal space separation (tSSS) method [21] with a subspace correlation limit of 0.9 [22,23] using MaxFilter software (Elekta Neuromag, version 2.2). Additionally, the automatic detection of saturated and bad MEG channels was also applied in the software. The bad channels detected were excluded from tSSS analysis to prevent artefacts spreading. The resultant MEG data were analysed with MATLAB R2019a using the Fieldtrip [24] and Brainstorm [25] toolboxes. The raw MEG data was filtered between 1–100 Hz and a bandstop filter of 48–52 Hz was also applied before recordings were resampled to 300 Hz. Independent Component Analysis (ICA) was used to decompose the MEG data, identify and subsequently remove eye-blink and cardiac artefacts. The components related to eye-blink and cardiac activity were identified by comparing the ICA component with the EOG and ECG recordings.

The power spectra were estimated using Welch's method with a Hanning window of 3 s with a 50% overlap. The relevant epochs were then extracted for each working memory load condition and power spectral density (PSD) estimates averaged across all MEG channels. PSDs were then normalized by dividing by the integral power between 1 Hz and 50 Hz to control for inherent differences within each participant and the average power spectra binned to the frequency of interest-gamma band activity (30–45Hz).

The implanted DRG stimulators used were not MRI compatible and, as such, individual structural MRIs (pre- or post-operative) were not available. Therefore, the ICB152 MRI template in Brainstorm was warped to fit the head model of each participant by co-registering the nasion, left and right pre-auricular fixed points acquired during head shape digitization [26]. Each subject-specific template was then used to calculate a lead field matrix based on a single shell model. The subsequent head model was co-registered with MEG data, and source localization performed using the dynamical imaging of coherent sources (DICS) beamformer technique based on the frequency of interest (30–45 Hz) of the processed MEG signals.

2.6. Statistical Analysis

Statistical analyses of MEG data to determine normalized PSD differences between *ON* and *OFF* stimulation was based on the non-parametric cluster-based permutation tests in the Fieldtrip toolbox [27]. A cluster was defined as two or more adjacent sensors reaching the pre-determined level of significance (t-statistic < 0.05). Statistical significance determined using the Monte Carlo method (p-value < 0.05, two-tailed) in order to correct for multiple comparisons. Comparisons of relative power between resting state and task performance conditions were calculated by finding the difference in the relative power between the two conditions and normalizing to the baseline power of the resting state condition to correct for inter-subject variability. The GraphPad Prism software version 8.1 (La Jolla California, CA, USA, www.graphpad.com) was used for other figures and statistical analyses presented. D'Agostino normality testing was conducted on each data set to confirm Gaussian distribution and the corresponding parametric test — Student's t-test or mixed-effects ANOVA (for comparisons of three or more groups) were utilized for analyses, respectively. *P*-values < 0.05 were regarded as statistically significant.

2.7. Mediation Analysis

A two-tailed Pearson correlation was performed to identify the relationship between gamma-band activity and patients' reported pain scores and task reaction times. Mediation analysis was conducted using SPSS (version 26) to assess whether there was a mediating effect between pain-related and cognition-related gamma activity in the frontal cortex, somatosensory cortex and dorsolateral prefrontal cortex. Mediation was tested by means of the joint significance test [28].

3. Results

Sixteen participants were recruited (10 males, 6 females) with an average age of 51 years (SD 16.5). However, only thirteen patients were included in MEG analysis after excluding data with unacceptable artefact/missing MEG channels. Contrary to expectation, only three of the sixteen participants reported alleviation of pain during task performance during the *DRGS-OFF* condition. The majority reported either worsening of pain scores ($n = 8$), or no change in pain ($n = 3$) during task performance compared to rest (see Figure 3). Interestingly, our cohort also included patients with posture-dependent/mobility-associated chronic pain syndromes ($n = 2$), which meant they did not report any pain at rest or during the task performance.

However, there was a significant reduction in reported pain scores (mean reduction 23% (SD 0.27), (F(2,30) = 10.33, $p = 0.001$) when DRGS was switched *ON* during the task, compared to *DRGS-OFF* during rest ($p = 0.01$) and task conditions ($p = 0.005$) (See Figure 3).

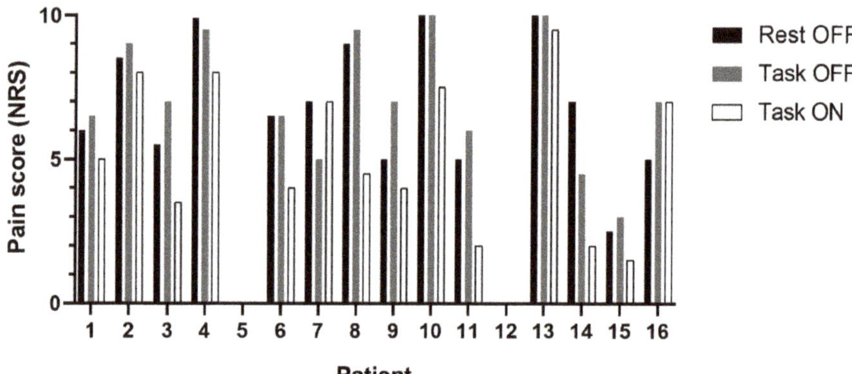

Figure 3. Grouped column graph depicting change from baseline pain scores at rest (black) and during n-back task performance (grey) with dorsal root ganglion stimulators (DRGS) turned off, as well as during task performance with DRGS turned on (white) among the sixteen participants. Of note, patients 5 and 12 had mobility-associated/posture-dependent pain and served as a unique "no-pain control" for the study.

3.1. Task Performance

There was a significant reduction in task accuracy ($F_{(2,24)}$ = 36.25, $p < 0.0001$) (See Figure 4A) and prolongation of RT ($F_{(2,24)}$ = 14.59, $p < 0.0001$) (See Figure 4B) in response to increasing attentional loads. There was no significant difference in RTs between 0-back and 1-back conditions, regardless of stimulation condition (OFF stimulation, $p = 0.98$, ON stimulation $p = 0.73$). However, the effect of working memory load on RT was driven by differences between the two lower working memory loads (0-back/1-back) and high working memory load (2-back) for both OFF ($p < 0.001$) and ON ($p = 0.004$) stimulation conditions (See Figure 4B).

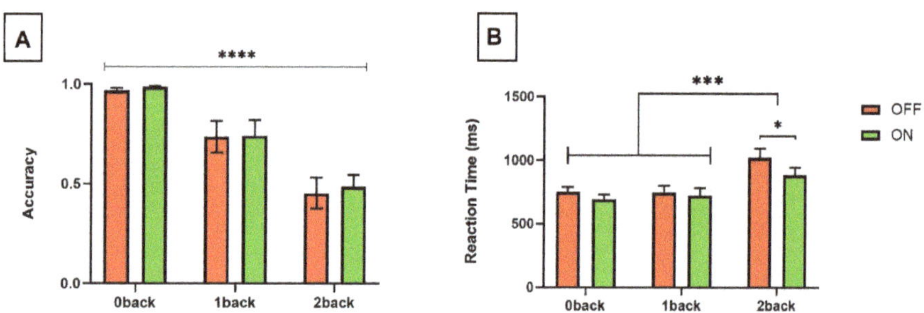

Figure 4. Bar graphs illustrating (**A**) task accuracy (proportion of correctly identified hits of all targets presented) and (**B**) reaction time with DRGS OFF (red) and ON (green) over increasing working memory loads. $p < 0.0001$ - ****; $p < 0.001$ - ***; $p < 0.05$ - *.

DRG stimulation was associated with a significant reduction in reaction time ($F_{(1,12)}$ = 6.516, $p = 0.025$), with posthoc tests confirming the statistical difference within the highest working memory load (2-back) condition ($p = 0.011$) (See Figure 4B). In contrast, there was no significant difference in task accuracy in response to DRGS across any working memory load condition ($F_{(1,12)}$ = 0.722, $p = 0.41$) (See Figure 4A).

3.2. Gamma Band Activity

Of the patients included in the MEG analysis experiencing pain during the study ($n = 11$), five reported 50% or greater reduction in reported pain scores with DRGS, while one reported worsening of pain. DRGS-mediated pain relief was associated with a significant reduction in gamma activity (30–45 Hz) across all MEG sensors during task performance ($t = 2.27$, $p = 0.036$) (See Figure 5A). The observed reduction in gamma band activity during pain relief was predominantly localized to the prefrontal cortex based on source-space analyses, but also revealed reductions in gamma activity in both somatosensory and anterior cingulate cortices after 3D source reconstruction (See Figure 5B).

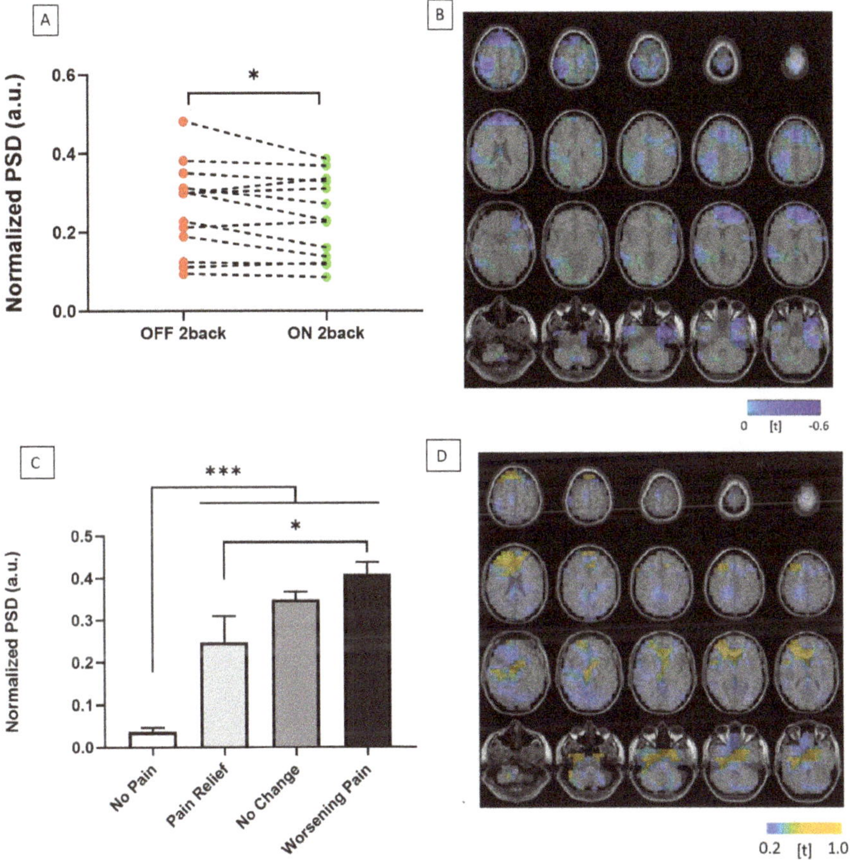

Figure 5. (**A**) Graph illustrating change in normalized power spectral density (PSD) between OFF (red) and ON (green) DRGS during high cognitive load (2-back condition). (**B**) 3-D source localization demonstrating t-statistic maps of significant reductions in gamma cortical activity across the prefrontal, anterior cingulate and somatosensory cortices during DRGS-mediated pain relief. (**C**) Column graph illustrating change in normalized power spectral density (PSD) with DRGS OFF, during high working memory load (2-back condition) compared to resting-state, grouped according to pain response during working memory load: no pain ($n = 2$), pain relief ($n = 2$), no change ($n = 3$) and worsening pain ($n = 6$) groups (A total of 13 patients were included in the MEG analysis). (**D**) 3-D source localization demonstrating t-maps, as before, of significant increases in cortical activity across the prefrontal and anterior cingulate cortices during task performance.

There were significant differences in gamma band fluctuations, dependent on the interaction of distraction and pain scores (F(3,80) = 65.01, $p < 0.001$). All groups exhibited increased gamma oscillatory activity during task performance compared to resting state. There was significantly greater gamma activity during task performance among those patients experiencing pain compared to pain-free controls ($p < 0.001$) (see Figure 6). Furthermore, among those in the pain-state, there was a significantly greater change in gamma oscillatory activity in patients that reported worsening pain during the task compared to those that exhibited pain relief during the attentional task ($p = 0.01$) (See Figure 5C). This increased gamma activity was also localized to the prefrontal and anterior cingulate cortices (See Figure 5D).

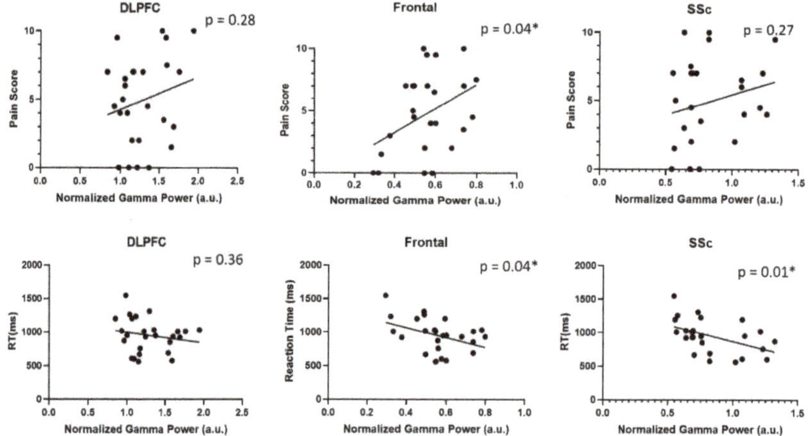

Figure 6. Graphs depicting correlations between reported pain scores and normalized gamma activity (top row), as well as correlations between reaction time and 2-back reaction times (bottom row) in the dorsolateral prefrontal cortex, frontal cortex and somatosensory cortex (SSc).

A significant correlation was found between gamma-band activity and subjectively reported pain scores in the frontal cortex([$r = 0.4$, $p = 0.04$). Additionally, significant correlations were observed between gamma-band activity and reaction times in both frontal and somatosensory cortices (See Figure 6). However, further analysis did not reveal a mediating effect of pain on cognition, or vice-versa (See Table 2).

Table 2. Mediation effects between pain-related gamma activity and cognition-related gamma activity in frontal, somatosensory and dorsolateral prefrontal cortices.

	Standardized β	Standard Error	p-value
Frontal			
Pain → Gamma	0.398	0.008	0.044
Cognition → Gamma	−0.332	0.00	0.082
Somatosensoy cortex			
Pain → Gamma	0.93	0.014	0.63
Cognition → Gamma	−0.447	0.00	0.028
Dorsolateral Prefrontal cortex			
Pain → Gamma	0.179	0.019	0.4
Cognition → Gamma	−0.134	0.00	0.53

4. Discussion

Our findings demonstrate the efficacy of DRGS in alleviating the interruptive effect of pain on cognition and supports the use of neurophysiologic signals, in particular, gamma-band activity, to interrogate the cognitive dimension of pain. We further demonstrate that while increased cognitive load is reflected by enhanced gamma oscillatory activity, the effect of pain, and pain relief, can modulate gamma activity in the human prefrontal and anterior cingulate cortices. Furthermore, our findings demonstrate that while frontal gamma activity was correlated with pain and cognitive measures, there was no mediating effect of pain on cognition, or vice-versa, which suggests that the potential for pain and cognition to modulate cortical gamma activity occur independently.

An inverse relationship is to be expected between task accuracy and reaction time with increasing cognitive load [29]. Accordingly, the n-back task results in our chronic pain cohort showed a significant reduction in task accuracy and a concomitant increase in reaction times with increasing working memory loads. However, cognitive loading (working memory) failed to alleviate pain in the majority of our participants. The phenomenon of distraction-induced analgesia is equivocal, having demonstrated mixed results across the pain literature. While there are studies which suggest that selective attention can mitigate the sensation of pain [14,30], there are also studies which have found that distraction can also exacerbate the perception of pain [31], as was seen in seven of the sixteen participants recruited in this study. Interestingly, the studies which demonstrate the phenomenon of distraction-mediated analgesia have been performed in healthy adults with the application of experimentally-induced pain. However, the initial report of worsened post-distraction pain [31], was performed in a cohort of chronic back pain patients which, taken together with our findings, suggests that this mechanism of pain alleviation may not be as applicable in chronic pain as previously thought.

It is classically believed that attention processing has a limited capacity, and by re-directing a portion of attentional reserves towards a cognitively demanding exercise, such as the n-back task, the accessibility of pain processing to this attentional network is decreased [32–34]. However, this mechanism of attentional switching seems to be sensitive to the degree of pain and the demands of the task on central attention [35,36]. A pleasant, moderately-engaging task might produce the intended alleviation of the pain percept by gating the accessibility of salient noxious stimuli to conscious processing. However, it seems similarly plausible that the challenge of a difficult, cognitively-demanding task can become frustrating and potentially exacerbate pain perception.

The disruptive effect of pain on task performance (accuracy) was not found to be significant in our cohort, despite marginal increases in accuracy during therapeutic DRGS. However, participants' reaction times were significantly reduced for a given level of accuracy, particularly in the high working memory load (2-back) condition. This suggests that with the alleviation of chronic pain, reduced response latency can be achieved without sacrificing task performance. Pain is a well-known interruptive factor in cognitive performance [37–39], and, persons suffering from chronic pain have been shown to exhibit deficits in various aspects of cognitive function including attention and memory [40,41]. The impact of pain on cognition seems to be dependent on the attentional load required of the task [42,43], which has similarly been demonstrated by our findings. The majority of these studies have been conducted with experimentally-induced pain in healthy adult participants. However, our study benefited from the ability to investigate the effect of acute pain relief, through neuromodulation, within the chronic pain phenotype and demonstrated its ability to improve performance on a cognitive task.

Our findings are bolstered by incorporating a well-established neurophysiologic signature, gamma oscillatory activity, as an objective metric of pain and attention. The neurophysiologic importance of gamma-band activity in the attentional modulation of pain has been previously demonstrated in healthy controls [44]. While our findings support the academic consensus which describes increased gamma activity in response to increased attentional demands [45,46], we further delineate the potential for pain to modulate this gamma activity.

DRGS-induced pain relief was associated with significantly reduced gamma activity during task performance (See Figure 5A). While a previous MEG study of SCS has hypothesized about the potential for increased cortical gamma activity in chronic pain [47], our findings have provided further support for this proposed mechanism of thalamocortical dysrhythmia. Our study also benefitted from a "no-pain control" group in this chronic pain cohort. Interestingly, in the *DRGS-OFF* condition, the "no-change" and "no-pain" groups also showed a significant disparity in gamma activity despite neither group having reported benefit from distraction-mediated analgesia (See Figure 5C). This observation suggests that this increased gamma activity is representative of ongoing pain in the chronic pain cortical network of the "no-change" group. Furthermore, we observed significantly lower gamma activity among participants reporting pain relief during task performance, compared to those reporting worsening pain (See Figure 5C). Taken altogether, our results suggest that the blunted increase in gamma activity we observed during task performance is likely a consequence of pain alleviation from distraction. However, it is also possible that pain relief in this group occurred in response to distraction-mediated analgesia, and this dampened gamma activity may represent the diversion of limited attentional resources. Further studies are required to conclusively disambiguate the causal relationship between these two possibilities.

The results of MEG source localization revealed gamma activation in brain regions which are known to be involved in the overlapping network of pain and attention, including somatosensory cortex [48,49] and cingulate cortices [50,51]. However, the observed changes in gamma activity were predominantly localized to the prefrontal cortex, which has been implicated in the top-down attentional modulation of painful stimuli [52] and has also been identified as a region that encodes ongoing pain among chronic pain patients and healthy adults [53,54]. Similar findings of attenuated cortical activity in cortico-limbic networks during DRG stimulation has been demonstrated in pre-clinical studies [55] and EEG studies of SCS [56]. Coupled with our findings of increased gamma activity during cognitive loads, and decreased gamma activity during pain relief in the prefrontal cortex, this represents further supportive evidence of the supraspinal effects of DRG stimulation.

The authors acknowledge the study limitations of a small sample size, resulting from the novelty of DRGS as an intervention for chronic pain. However, the utilization of a crossover study design was employed to overcome this limitation and increase statistical power by minimizing between-subject variability. We also recognize that such an overlap in cortical networks between pain-related and attention-related activities may still be represented by more functionally distinct anatomical regions than the areas described in our analysis. Further elucidation of these anatomical differences might be achieved by combining techniques such as fMRI which can resolve deeper anatomical structures involved in the pain network (insular cortex, thalamus) with greater sensitivity and spatial resolution. These limitations notwithstanding, this study offers novel evidence for the supraspinal effects of DRGS in chronic pain and demonstrates the importance of gamma oscillatory activity in the neurophysiologic representation of pain and cognition.

Author Contributions: T.P.–Conception of experimental design, conducted experiments, data analysis, drafted original manuscript, edit/review of final manuscript. Y.H.–Data analysis, edit/review of final manuscript. A.L.B.R.–Edit/review of final manuscript. J.J.F.-Project supervision, edit/review of final manuscript. A.L.G.–Project supervision, conception of experimental design, edit/review of final manuscript. T.Z.A.-Project supervision, edit/review of final manuscript. All authors have read and agreed to the published version of the manuscript.

Funding: This research was funded by the Oxford Hospitals Charity (Registered charity number 1175809). T.P. is also supported by the Howard Brain Sciences Foundation and the Rhodes Trust.

Acknowledgments: The authors wish to acknowledge Sven Braetigum and Rocio Silva Zunino for their dedication and assistance in conducting MEG recordings. We also appreciate the contributions and support of B.A.P and all the researchers and staff at the Oxford Centre for Human Brain Activity. Most importantly, the authors are grateful to the patients that volunteered to participate in this study, without whom this work would not be possible.

Conflicts of Interest: Fitzgerald reports being a consultant to Abbott and Medtronic, and Green reports being a consultant to Abbott and Herantis Pharma Plc. There are no other relevant conflicts of interest to disclose.

References

1. Melzack, R.; Casey, K. Sensory, Motivational, and Central Control Determinants of Pain. In *The Skin Senses: Proceedings*; Thomas, C.C., Ed.; Springfield: Illinois, IL, USA, 1968; pp. 423–439.
2. Bushnell, M.C.; Čeko, M.; Low, L.A. Cognitive and emotional control of pain and its disruption in chronic pain. *Nat. Rev. Neurosci.* **2013**, *14*, 502–511. [CrossRef] [PubMed]
3. Torta, D.M.; Legrain, V.; Mouraux, A.; Valentini, E. Attention to pain! A neurocognitive perspective on attentional modulation of pain in neuroimaging studies. *Cortex* **2017**, *89*, 120–134. [CrossRef] [PubMed]
4. Schreiber, K.L.; Campbell, C.; Martel, M.O.; Greenbaum, S.; Wasan, A.D.; Borsook, D.; Jamison, R.N.; Edwards, R.R. Distraction analgesia in chronic pain patients the impact of catastrophizing. *Anesthesiology* **2014**, *121*, 1292–1301. [CrossRef] [PubMed]
5. Valet, M.; Sprenger, T.; Boecker, H.; Willoch, F.; Rummeny, E.; Conrad, B.; Erhard, P.; Tolle, T.R. Distraction modulates connectivity of the cingulo-frontal cortex and the midbrain during pain—An fMRI analysis. *Pain* **2004**, *109*, 399–408. [CrossRef]
6. Berryman, C.; Stanton, T.R.; Jane Bowering, K.; Tabor, A.; McFarlane, A.; Lorimer Moseley, G. Evidence for working memory deficits in chronic pain: A systematic review and meta-analysis. *Pain* **2013**, *154*, 1181–1196. [CrossRef]
7. Eccleston, C.; Crombez, G. Pain demands attention: A cognitive-affective model of the interruptive function of pain. *Psychol. Bull.* **1999**, *125*, 356–366. [CrossRef]
8. Gruber, T.; Müller, M.M.; Keil, A.; Elbert, T. Selective visual-spatial attention alters induced gamma band responses in the human EEG. *Clin. Neurophysiol.* **1999**, *110*, 2074–2085. [CrossRef]
9. Tallon-Baudry, C.; Bertrand, O.; Hénaff, M.-A.; Isnard, J.; Fischer, C. Attention Modulates Gamma-band Oscillations Differently in the Human Lateral Occipital Cortex and Fusiform Gyrus. *Cereb. Cortex* **2005**, *15*, 654–662. [CrossRef]
10. De Pascalis, V.; Cacace, I.; Massicolle, F. Perception and modulation of pain in waking and hypnosis: Functional significance of phase-ordered gamma oscillations. *Pain* **2004**, *112*, 27–36. [CrossRef]
11. Tan, L.L.; Oswald, M.J.; Heinl, C.; Retana Romero, O.A.; Kaushalya, S.K.; Monyer, H.; Kuner, R. Gamma oscillations in somatosensory cortex recruit prefrontal and descending serotonergic pathways in aversion and nociception. *Nat. Commun.* **2019**, *10*, 983. [CrossRef]
12. Bai, Y.; Xia, X.; Liang, Z.; Wang, Y.; Yang, Y.; He, J.; Li, X. Frontal Connectivity in EEG Gamma (30–45 Hz) Respond to Spinal Cord Stimulation in Minimally Conscious State Patients. *Front. Cell. Neurosci.* **2017**, *11*, 177. [CrossRef] [PubMed]
13. Deer, T.R.; Levy, R.M.; Kramer, J.; Poree, L.; Amirdelfan, K.; Grigsby, E.; Staats, P.; Burton, A.W.; Burgher, A.H.; Obray, J.; et al. Dorsal root ganglion stimulation yielded higher treatment success rate for complex regional pain syndrome and causalgia at 3 and 12 months: A randomized comparative trial. *Pain* **2017**, *158*, 669–681. [CrossRef] [PubMed]
14. Bantick, S.J.; Wise, R.G.; Ploghaus, A.; Clare, S.; Smith, S.M.; Tracey, I. Imaging how attention modulates pain in humans using functional MRI. *Brain* **2002**, *125*, 310–319. [CrossRef] [PubMed]
15. Tracey, I.; Ploghaus, A.; Gati, J.S.; Clare, S.; Smith, S.; Menon, R.S.; Matthews, P.M. Imaging Attentional Modulation of Pain in the Periaqueductal Gray in Humans. *J. Neurosci.* **2002**, *22*, 2748–2752. [CrossRef]
16. Seminowicz, D.A.; Davis, K.D. Interactions of pain intensity and cognitive load: The brain stays on task. *Cereb. Cortex* **2007**, *17*, 1412–1422. [CrossRef]
17. Sprenger, C.; Eippert, F.; Finsterbusch, J.; Bingel, U.; Rose, M.; Büchel, C. Attention Modulates Spinal Cord Responses to Pain. *Curr. Biol.* **2012**, *22*, 1019–1022. [CrossRef]
18. Moore, D.J.; Keogh, E.; Eccleston, C. The effect of threat on attentional interruption by pain. *Pain* **2013**, *154*, 82–88. [CrossRef] [PubMed]
19. Attridge, N.; Noonan, D.; Eccleston, C.; Keogh, E. The disruptive effects of pain on n-back task performance in a large general population sample. *Pain* **2015**, *156*, 1885–1891. [CrossRef]
20. Parker, T.; Green, A.; Aziz, T. Rapid onset and short washout periods of dorsal root ganglion stimulation facilitate multiphase crossover study designs. *Brain Stimul.* **2019**, *12*, 1617–1618. [CrossRef]
21. Taulu, S.; Simola, J. Spatiotemporal signal space separation method for rejecting nearby interference in MEG measurements. *Phys. Med. Biol.* **2006**, *51*, 1759–1768. [CrossRef]

22. Medvedovsky, M.; Taulu, S.; Bikmullina, R.; Ahonen, A.; Paetau, R. Fine tuning the correlation limit of spatio-temporal signal space separation for magnetoencephalography. *J. Neurosci. Methods* **2009**, *177*, 203–211. [CrossRef] [PubMed]

23. Carrette, E.; De Tiège, X.; Op De Beeck, M.; De Herdt, V.; Meurs, A.; Legros, B.; Raedt, R.; Deblaere, K.; Van Roost, D.; Bourguignon, M.; et al. Magnetoencephalography in epilepsy patients carrying a vagus nerve stimulator. *Epilepsy Res.* **2011**, *93*, 44–52. [CrossRef] [PubMed]

24. Oostenveld, R.; Fries, P.; Maris, E.; Schoffelen, J.M. FieldTrip: Open source software for advanced analysis of MEG, EEG, and invasive electrophysiological data. *Comput. Intell. Neurosci.* **2011**. [CrossRef] [PubMed]

25. Tadel, F.; Baillet, S.; Mosher, J.C.; Pantazis, D.; Leahy, R.M. Brainstorm: A user-friendly application for MEG/EEG analysis. *Comput. Intell. Neurosci.* **2011**, 879716. [CrossRef]

26. Gohel, B.; Lim, S.; Kim, M.-Y.; Kwon, H.; Kim, K. Approximate Subject Specific Pseudo MRI from an Available MRI Dataset for MEG Source Imaging. *Front. Neuroinform.* **2017**, *11*, 50. [CrossRef]

27. Maris, E.; Oostenveld, R. Nonparametric statistical testing of EEG- and MEG-data. *J. Neurosci. Methods* **2007**, *164*, 177–190. [CrossRef]

28. MacKinnon, D.P.; Lockwood, C.M.; Hoffman, J.M.; West, S.G.; Sheets, V. A comparison of methods to test mediation and other intervening variable effects. *Psychol. Methods* **2002**, *7*, 83–104. [CrossRef]

29. Meule, A. Reporting and interpreting task performance in Go/no-go affective shifting tasks. *Front. Psychol.* **2017**, *8*, 701. [CrossRef]

30. Buhle, J.; Wager, T.D. Performance-dependent inhibition of pain by an executive working memory task. *Pain* **2010**, *149*, 19–26. [CrossRef]

31. Goubert, L.; Crombez, G.; Eccleston, C.; Devulder, J. Distraction from chronic pain during a pain-inducing activity is associated with greater post-activity pain. *Pain* **2004**, *110*, 220–227. [CrossRef]

32. McCaul, K.D.; Malott, J.M. Distraction and coping with pain. *Psychol. Bull.* **1984**, *95*, 516–533. [CrossRef] [PubMed]

33. Kahneman, D. *Attention and Effort*; Prentice-Hall: Englewood Cliffs, NJ, USA, 1973.

34. Broadbent, D. *Perception and Communication*; Elsevier: Amsterdam, The Netherlands, 1958.

35. Eccleston, C. Chronic pain and distraction: An experimental investigation into the role of sustained and shifting attention in the processing of chronic persistent pain. *Behav. Res. Ther.* **1995**, *33*, 391–405. [CrossRef]

36. Roa Romero, Y.; Straube, T.; Nitsch, A.; Miltner, W.H.R.; Weiss, T. Interaction between stimulus intensity and perceptual load in the attentional control of pain. *Pain* **2013**, *154*, 135–140. [CrossRef] [PubMed]

37. Moore, D.J.; Keogh, E.; Eccleston, C. Headache impairs attentional performance. *Pain* **2013**, *154*, 1840–1845. [CrossRef]

38. Keogh, E.; Cavill, R.; Moore, D.J.; Eccleston, C. The effects of menstrual-related pain on attentional interference. *Pain* **2014**, *155*, 821–827. [CrossRef]

39. Attridge, N.; Keogh, E.; Eccleston, C. The effect of pain on task switching: Pain reduces accuracy and increases reaction times across multiple switching paradigms. *Pain* **2016**, *157*, 2179–2193. [CrossRef]

40. Nadar, M.S.; Jasem, Z.; Manee, F.S. The cognitive functions in adults with chronic pain: A comparative study. *Pain Res. Manag.* **2016**, *2016*, 5719380. [CrossRef]

41. Moriarty, O.; Ruane, N.; O'Gorman, D.; Maharaj, C.H.; Mitchell, C.; Sarma, K.M.; Finn, D.P.; McGuire, B.E. Cognitive impairment in patients with chronic neuropathic or radicular pain: An interaction of pain and age. *Front. Behav. Neurosci.* **2017**, *11*, 100. [CrossRef]

42. Moore, D.J.; Keogh, E.; Eccleston, C. The Interruptive Effect of Pain on Attention. *Q. J. Exp. Psychol.* **2012**, *65*, 565–586. [CrossRef]

43. Moore, D.J.; Eccleston, C.; Keogh, E. Cognitive load selectively influences the interruptive effect of pain on attention. *Pain* **2017**, *158*, 2035–2041. [CrossRef]

44. Hauck, M.; Lorenz, J.; Engel, A.K. Attention to painful stimulation enhances γ-band activity and synchronization in human sensorimotor cortex. *J. Neurosci.* **2007**, *27*, 9270–9277. [CrossRef]

45. Jensen, O.; Kaiser, J.; Lachaux, J.P. Human gamma-frequency oscillations associated with attention and memory. *Trends Neurosci.* **2007**, *30*, 317–324. [CrossRef] [PubMed]

46. Ray, S.; Niebur, E.; Hsiao, S.S.; Sinai, A.; Crone, N.E. High-frequency gamma activity (80–150 Hz) is increased in human cortex during selective attention. *Clin. Neurophysiol.* **2008**, *119*, 116–133. [CrossRef] [PubMed]

47. Schulman, J.J.; Ramirez, R.R.; Zonenshayn, M.; Ribary, U.; Llinas, R. Thalamocortical dysrhythmia syndrome: MEG imaging of neuropathic pain. *Thalamus Relat. Syst.* **2005**, *3*, 33–39. [CrossRef]

48. Petrovic, P.; Ingvar, M.; Stone-Elander, S.; Petersson, K.M.; Hansson, P. A PET activation study of dynamic mechanical allodynia in patients with mononeuropathy. *Pain* **1999**, *83*, 459–470. [CrossRef]

49. Petrovic, P.; Petersson, K.M.; Ghatan, P.H.; Stone-Elander, S.; Ingvar, M. Pain-related cerebral activation is altered by a distracting cognitive task. *Pain* **2000**, *85*, 19–30. [CrossRef]

50. Davis, K.D.; Taylor, S.J.; Crawley, A.P.; Wood, M.L.; Mikulis, D.J. Functional MRI of pain- and attention-related activations in the human cingulate cortex. *J. Neurophysiol.* **1997**, *77*, 3370–3380. [CrossRef]

51. Schmidt-Wilcke, T.; Kairys, A.; Ichesco, E.; Fernandez-Sanchez, M.L.; Barjola, P.; Heitzeg, M.; Harris, R.E.; Clauw, D.J.; Glass, J.; Williams, D.A. Changes in clinical pain in fibromyalgia patients correlate with changes in brain activation in the cingulate cortex in a response inhibition task. *Pain Med. (USA)* **2014**, *15*, 1346–1358. [CrossRef]

52. Legrain, V.; Van Damme, S.; Eccleston, C.; Davis, K.D.; Seminowicz, D.A.; Crombez, G. A neurocognitive model of attention to pain: Behavioral and neuroimaging evidence. *Pain* **2009**, *144*, 230–232. [CrossRef]

53. Schulz, E.; May, E.S.; Postorino, M.; Tiemann, L.; Nickel, M.M.; Witkovsky, V.; Schmidt, P.; Gross, J.; Ploner, M. Prefrontal Gamma Oscillations Encode Tonic Pain in Humans. *Cereb. Cortex* **2015**, *25*, 4407–4414. [CrossRef]

54. May, E.S.; Nickel, M.M.; Ta Dinh, S.; Tiemann, L.; Heitmann, H.; Voth, I.; Tölle, T.R.; Gross, J.; Ploner, M. Prefrontal gamma oscillations reflect ongoing pain intensity in chronic back pain patients. *Hum. Brain Mapp.* **2019**, *40*, 293–305. [CrossRef] [PubMed]

55. Pawela, C.P.; Kramer, J.M.; Hogan, Q.H. Dorsal root ganglion stimulation attenuates the BOLD signal response to noxious sensory input in specific brain regions: Insights into a possible mechanism for analgesia. *Neuroimage* **2017**, *147*, 10–18. [CrossRef] [PubMed]

56. De Ridder, D.; Plazier, M.; Kamerling, N.; Menovsky, T.; Vanneste, S. Burst spinal cord stimulation for limb and back pain. *World Neurosurg.* **2013**, *80*, 642–649. [CrossRef] [PubMed]

Article

Manual Therapy Reduces Pain Behavior and Oxidative Stress in a Murine Model of Complex Regional Pain Syndrome Type I

Afonso S. I. Salgado [1,2], Juliana Stramosk [2], Daniela D. Ludtke [2,3], Ana C. C. Kuci [2], Daiana C. Salm [2,3], Lisandro A. Ceci [2,3], Fabricia Petronilho [4], Drielly Florentino [4], Lucineia G. Danielski [4], Aline Gassenferth [5], Luana R. Souza [6], Gislaine T. Rezin [7], Adair R. S. Santos [5,6], Leidiane Mazzardo-Martins [6], William R. Reed [7] and Daniel F. Martins [1,2,*]

[1] Coordinator of Integrative Physical Therapy Residency–Philadelphia University Center, Londrina 86020-000, Paraná, Brazil

[2] Experimental Neuroscience Laboratory (LaNEx), University of Southern Santa Catarina, Palhoça 88137-270, Santa Catarina, Brazil

[3] Postgraduate Program in Health Sciences, University of Southern Santa Catarina, Palhoça 88137-270, Santa Catarina, Brazil

[4] Laboratory of Neurobiology of Inflammatory and Metabolic Processes, Graduate Program in Health Sciences, Health Sciences Unit, University of South Santa Catarina, Tubarão 88704-900, Santa Catarina, Brazil

[5] Laboratory of Neurobiology of Pain and Inflammation, Department of Physiological Sciences, Centre of Biological Sciences, University Federal of Santa Catarina, Florianópolis 88040-900, Santa Catarina, Brazil

[6] Postgraduate Program in Neuroscience, Center of Biological Sciences, Federal University of Santa Catarina, Florianópolis 88040-900, Santa Catarina, Brazil

[7] Department of Physical Therapy, School of Health Professions, University of Alabama at Birmingham, Birmingham, AL 35294-1212, USA

* Correspondence: daniel.martins4@unisul.br; Tel.: +55-(48)-3279-1144

Received: 2 July 2019; Accepted: 8 August 2019; Published: 10 August 2019

Abstract: Complex regional pain syndrome type I (CRPS-I) is a chronic painful condition. We investigated whether manual therapy (MT), in a chronic post-ischemia pain (CPIP) model, is capable of reducing pain behavior and oxidative stress. Male Swiss mice were subjected to ischemia-reperfusion (IR) to mimic CRPS-I. Animals received ankle joint mobilization 48h after the IR procedure, and response to mechanical stimuli was evaluated. For biochemical analyses, mitochondrial function as well as oxidative stress thiobarbituric acid reactive substances (TBARS), protein carbonyls, antioxidant enzymes superoxide dismutase (SOD) and catalase (CAT) levels were determined. IR induced mechanical hyperalgesia which was subsequently reduced by acute MT treatment. The concentrations of oxidative stress parameters were increased following IR with MT treatment preventing these increases in malondialdehyde (MDA) and carbonyls protein. IR diminished the levels of SOD and CAT activity and MT treatment prevented this decrease in CAT but not in SOD activity. IR also diminished mitochondrial complex activity, and MT treatment was ineffective in preventing this decrease. In conclusion, repeated sessions of MT resulted in antihyperalgesic effects mediated, at least partially, through the prevention of an increase of MDA and protein carbonyls levels and an improvement in the antioxidant defense system.

Keywords: chronic pain; Complex Regional Pain Syndrome; manual therapy; osteopathy; oxidative stress

---

1. Introduction

Complex regional pain syndrome type I (CRPS-I) is a chronic painful condition that frequently develops after a deep tissue injury, such as a fracture or sprain, without nerve injury. CPRS-I is clinically characterized by a variety of sensory disturbances including allodynia, hyperalgesia, edema, vasomotor/sudomotor deregulation, skin and underlying tissue trophism modification. Symptoms typically begin in the distal part of the affected limb and spread to the unaffected or opposite limb [1–7].

The pathophysiology of this syndrome remains unclear, however inflammatory and neural mechanisms have been suggested as potential contributors. Both peripheral and central mechanisms are thought to play a prominent role; however, evidence exists indicating that oxidative stress (OS) also plays an important role [5–7]. Individuals with CRPS-I suffer from alterations in central and peripheral nervous system processing leading to decreased pain pressure threshold and increased temporal summation of pain [8]. These physiological changes most likely involve OS changes, which are known to be an important mechanism following tissue injury and hypoxia [9,10].

A rodent model of chronic post-ischemia pain (CPIP) was developed by Coderre et al. [1] which mimics much of the clinical symptomatology associated with CRPS-I. This model was first developed in rats [1] and later adapted for mice [11]. It produces ischemia followed by reperfusion, and its initial phase is characterized by hyperemia and edema that produces micro vascular injury, deep tissue inflammation, muscle nociceptor activation and ectopic activation of afferent sensory axons via an inflammatory cascade and endoneural ischemia [1]. Reactive oxygen species (ROS) are known to play a predominate role in the inflammatory event cascade created by prolonged hindlimb ischemia-reperfusion (IR) [1,12,13] resulting in the production of oxidants, superoxide, hydroxyl radicals and hydrogen peroxide among others. Assuming that the generation of free radicals is partially responsible for CRPS-I in CPIP, Coderre et al. [1] demonstrated that two free radical eliminators reduced signs of mechanical allodynia emphasizing the importance of oxidants in the maintenance of CRPS-I neuropathic pain symptoms. Furthermore, the presence of OS in patients with CRPS-I patients has been indirectly confirmed thereby strengthening the rationale for clinical use of antioxidants and free radical scavengers to treat and/or prevent CRPS type I [14,15].

Pain management in combination with strength and flexibility training along with manual soft tissue techniques applied to the involved extremity have been the traditional clinical treatment of CRPS [16–18]. In this sense, typical treatment strategies include desensitization therapy, manual therapy, progressive exercise, and patient education [16–19]. Among conservative therapies, manual therapy (i.e., joint mobilization) stands out as a possible therapeutic for the reduction of symptoms and signs of CRPS-I since it is commonly used to treat a number of painful conditions [20]. Main clinical effects of manual therapy include pain reduction, functional improvement and aspects of neurophysiological modulation [21,22]. Manual therapy is commonly used to treat a variety of musculoskeletal conditions as an adjunct treatment, but literature describing its use for managing CRPS is scarce. Clinical and preclinical studies have provided a good rationale to test the effect of manual therapy (MT) on CRPS-I and to determine the physiological contribution of oxidative stress. For example, in a clinical case series of individuals experiencing bilateral lower extremity CRPS, application of MT to the lumbar spine along with traditional conservative care resulted in meaningful clinical outcomes that were most likely associated with the MT intervention [23]. Furthermore, Kolberg et al. [24] reported that joint manipulation in humans increased catalase (CAT) activity in erythrocytes showing the antioxidative effect of manual therapy intervention.

Our research group has demonstrated that activation of inhibitory neuroreceptors such as adenosine A, opioid, cannabinoid 2 (CB$_2$), peripheral/spinal and cannabinoid 1 (CB$_1$) are involved in analgesic effects of MT in mice [2,25,26]. Interestingly, these endogenous systems activated by MT modulate oxidative stress. In rats, stress-activation of lipid peroxidation in plasma and liver tissue was reduced by the injection of opioid peptides while at the same time increasing catalase activity [27]. In human monocytes/macrophages, it has been shown that during inflammation the CB$_1$ receptor is highly expressed and that its activation directly modulates inflammatory activity by means of production of ROS [28]. Moreover, the activation of the CB$_2$ receptor may generate inhibitory signaling that directly suppresses the production of ROS stimulated by the activation of the receptor CB$_1$ [29]. In addition, the adenosinergic system is known to modulate oxidative stress especially via activation of the A_1 receptor [30].

Although the neurophysiological effects of MT has been demonstrated in other animal pain models, to date it has not been investigated in a CRPS-I model. The purpose of this study was to determine if MT can indeed reduce pain behavior and oxidative stress by means of enzymatic anti-oxidative system activation in a CRPS-I model. Thus, the results of the present study may serve as a basis for future clinical trials aiming to evaluate the effects of MT on CRPS-I or other painful conditions that have oxidative stress as the main pathophysiology. In addition, this study also shows the possibility of beneficial effects in the association of MT with anti-oxidant therapies in the treatment of chronic pain.

2. Materials and Methods

2.1. Animals

All experimental procedures were approved by the Ethics Committee of the University of Southern Santa Catarina at Palhoça, Santa Catarina, Brazil (protocol number 15.034.3.07.IV) and performed in accordance with the National Institute of Health Animal Care Guidelines (NIH publications number 80-23). Male Swiss mice (25–35 g) were obtained from the Biotério Central da Universidade Federal de Santa Catarina (UFSC, Florianópolis, Santa Catarina, Brazil) and group housed at 22 ± 2 °C under a 12 h light/12 h dark cycle (lights on at 6 a.m.) with food and water ad libitum. Mice were habituated to the testing environment for a minimum of 1 h before any experiments were conducted between 8 a.m. and noon [31]. Figure 1 shows the experimental timeline of IR injury, MT treatment and tissue harvesting.

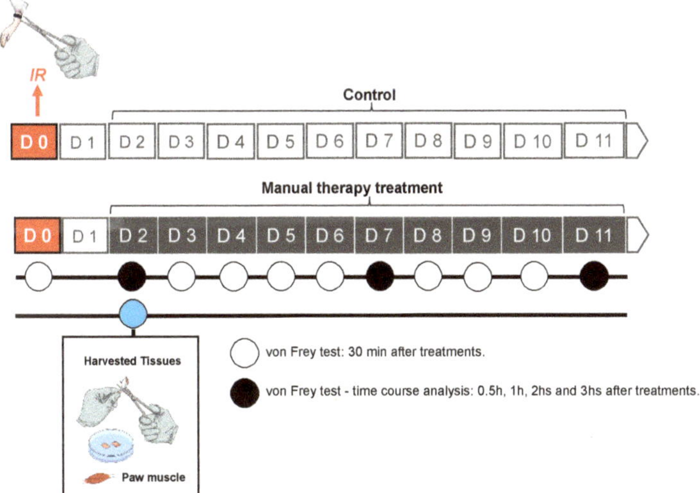

Figure 1. Timeline of treatment and analyses. Ischemia-reperfusion (IR): Ischemia and reperfusion; D: day; min: minutes; h: hour.

2.2. Animal Model CRPS-I

The animal model of CRPS-I was performed following experimental procedures described first for rats and later adapted for mice [11], involving exposure to prolonged hindpaw IR. This model uses an elastic O-ring (commonly used for orthodontic braces (Elástico Ligadura 000-1237, Uniden, SP, Brazil) with a 1.2-mm internal diameter placed around the right hindlimb just proximal to the ankle joint thereby producing ischemia. During this procedure, mice were anesthetized with a bolus (7%,

0.6 mL/kg, i.p.) of chloral hydrate and 20% of the initial volume at the end of the first and second hour. As previously established in rodent models, O-rings were left on the limb for 3 hours. All sham mice were subjected to the same experimental procedures except that the O-ring was slightly cut so that it only loosely surrounded the ankle so as to not occlude blood flow to the right hindpaw [32,33].

2.3. MT Treatment

MT treatment was performed as previously described [25]. Mice were lightly anesthetized with 1%–2% isoflurane and the experimenter's hand stabilized the knee joint while the ankle joint was flexed and extended to full amplitude, rhythmically with a movement frequency of approximately 40 cycles per minute. Movement frequency was performed with assistance of a metronome. Treated animals received a total of 9 minutes of MT divided in 3 series of 3 minutes each with a 30 second interval between series. Sham group animals were kept anesthetized for the same time period, with the experimenter's hands positioned on ankle joint but no movements were performed [2,26]. Animals received daily treatments of 9-minute MT between the 2nd to 11th day after the IR procedure.

2.4. Mechanical Hyperalgesia

To assess mechanical hyperalgesia, mice were acclimatized to individual clear boxes ($9 \times 7 \times 11$ cm) on an elevated wire mesh platform which allowed access to the ventral hindpaw surface, as previously described [2,25]. Mechanical hyperalgesia was measured with right hindpaw stimulation in a series of 10 non-consecutive applications using calibrated 0.4 g von Frey filaments (Stoelting, Chicago, IL, USA) [26]. Results are reported as the percentage of response frequency. The time course analyses of antihyperalgesic effects caused by MT was performed at the 2nd, 7th and 11th days after the IR procedure, at 30, 60 and 90 minutes after MT treatment. In a separate set of experiments, mechanical hyperalgesia was assessed every day following MT between the 2nd to 11th day after the IR procedure.

2.5. Sample Collection for Biochemical Analyses

In a separate set of experiments involving the collection of biological samples on the 2nd day after IR, all animals were euthanized 30 min after MT treatment and right hindpaw muscle tissue samples were surgically harvested. The tissues were weighed and homogenized in 10 volumes (1:10, *w/v*) of ice-cold 0.1 M phosphate buffer (pH 7.4). To discard cell debris and nuclei, homogenates were centrifuged at $750\times g$ for 10 min at 4 °C. After discarding the pellet, aliquots of supernatants were separated and used for determination of oxidative stress parameters.

2.6. Determination of Oxidative Stress and Antioxidant Enzymes Levels

Thiobarbituric acid reactive species (TBARS) formation was measured during an acid-heating reaction [34]. Samples were heated for 15min in a boiling water bath, mixed with 1ml of trichloroacetic acid (TCA) 10% and 1ml of thiobarbituric acid 0.67%. TBARS was determined by the absorbance at 535nm. Results are expressed as malondialdehyde (MDA) equivalents (nmol/mg protein).

Oxidative damage to proteins was measured by determining the carbonyl groups based on the reaction with dinitrophenylhydrazine (DNPH) [35]. Precipitation of proteins were conducted by the addition of 20% trichloroacetic acid and redissolved in DNPH with the absorbance read at 370 nm. Results were reported as nmol of carbonyl content per mg of protein (nmol/mg protein).

Catalase (CAT) activity was measured by the rate of decrease in hydrogen peroxide absorbance at 240 nm [36]. Briefly, hindpaw tissue samples were sonicated in 50nmol/l phosphate buffer (pH 7.0), and the resulting suspension was centrifuged at 3000 g for 10 min. The supernatant was used for the enzyme assay. Results were reported as units per milligram of protein (U/mg protein).

The activity of superoxide dismutase (SOD) was determined by measuring the inhibition of adrenaline auto-oxidation spectrophotometrically at 480 nm [37] and was represented as units per milligram of protein (U/mg protein).

Bovine albumin was used as a standard to normalize all biochemical measurements [38].

2.7. Mitochondrial Function Analyses

Nicotinamide adenine dinucleotide (NADH)-dependent ferricyanide reduction was used to measure Complex I activity [39]. Samples were coupled with reagents 100mM potassium phosphate buffer, 10 mM ferricyanide, 14 mM NADH and 2 mM rotenone, and analyzed at 420 nm by a spectrophotometer with readings taken minute by minute for a total of 3 minutes.

The 2,6-diclorophenolindophenol (DCPIP) reduction was used to measure Complex II activity as described by Fisher et al. [40]. The tissue sample was incubated for 20 minutes at 30 °C water bath with 62.5 mM potassium phosphate buffer, 250 mM sodium succinate and 0.5 mM DCPIP. After incubation, 100 mM sodium azide, 2 mM rotenone and 0.5 mM DCPIP were added and then minute by minute spectrophotometric readings taken for a total of 5 minutes at 600 nm.

Complex IV activity was determined by calculating the absorbance reduction caused by reduced cytochrome c oxidation as described by Rustin et al. [41]. In the incubation environment, 62.5 mM potassium phosphate buffer, 125 mM lauryl maltoside was added and sample diluted with SETH buffer (Sucrose, EDTA, Tris and Heparine) and 1% cytochrome c. Analyses were performed at 550 nm by a spectrophotometer with readings taken minute by minute for 10 min. The results of mitochondrial function were expressed in nmol/min by mg of protein.

2.8. Statistical Analysis

The Shapiro–Wilk test was used to evaluate the normality assumption of all behavioral and biochemical data. All variables in the present study showed a normal distribution. Differences among experimental groups were determined by one or two-way ANOVA followed by Student–Newman–Keuls Multiple Comparison or Bonferroni post hoc test, respectively, as appropriate. A value of $p < 0.05$ was considered to be statistically significant. All data are presented as mean (standard deviation, SD).

3. Results

Figure 2 shows that mice hindpaw IR induced marked and long-lasting mechanical hyperalgesia, as observed by the enhancement of the response frequency to the von Frey filament application in comparison to sham mice ($p = 0.001$) (Figure 2A–E). We observed that the acute MT treatment (IR + MT group) on the 2nd, 7th and 11th days after IR reduced mechanical hypersensitivity induced by IR. Significant differences between groups (IR + Control vs IR + MT) were observed at 0.5 h ($p = 0.001$) and 1 h ($p = 0.001$) after MT (Figure 2A,C,E). Furthermore, the repeated daily MT treatments (2–7 or 7–11 days) decreased ($p = 0.001$) the mechanical hypersensitivity induced by IR when assessed 30 minutes after MT (Figure 2B,D).

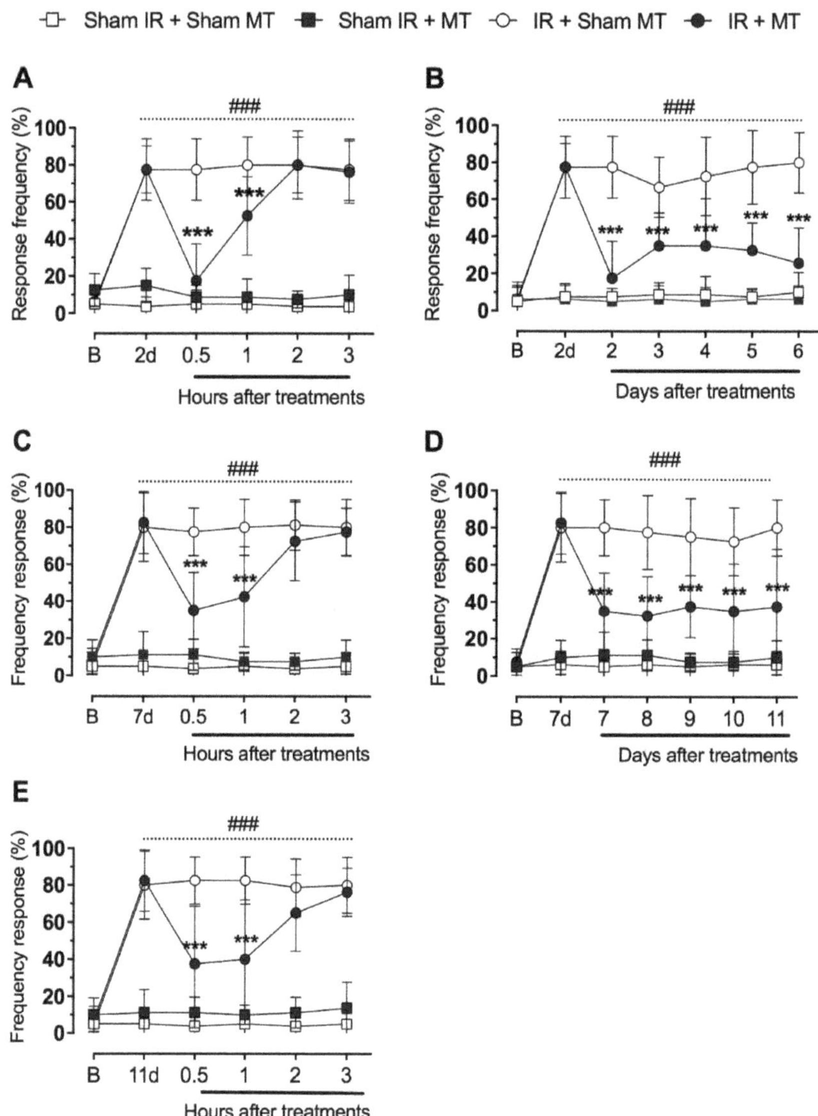

Figure 2. Effect of manual therapy (MT) on mechanical hyperalgesia. Time course analysis at the 2nd day (panel **A**). Daily treatment with 9-minute ankle joint mobilization between the 2nd to 6th day after IR procedure (panel **B**). Time course analysis at 7th day (panel **C**). Daily treatment with 9-minutes of MT between the 7th to 11th day after IR procedure (panel **D**). Time course analysis at 11th day (panel **E**). Each point represents the mean of 8 animals and vertical lines show the SD. Statistical analyses were performed by two-way ANOVA followed by Bonferroni test. The symbols denote a significant difference of *** $p < 0.001$ when compared to IR + Sham MT group or ### $p < 0.001$ when compared to Sham + Sham MT group. MT: Manual therapy, IR: Schemia and reperfusion.

Figure 3 shows that at the 2nd day after IR injury, the concentrations of MDA (Figure 3A, $p = 0.02$) and protein carbonyls (Figure 3B, $p = 0.01$) in muscle paw tissue were increased compared to Sham and Sham + MT groups. MT significantly prevented MDA (Figure 3A, $p = 0.03$) and carbonyls protein increase (Figure 3B, $p = 0.03$).

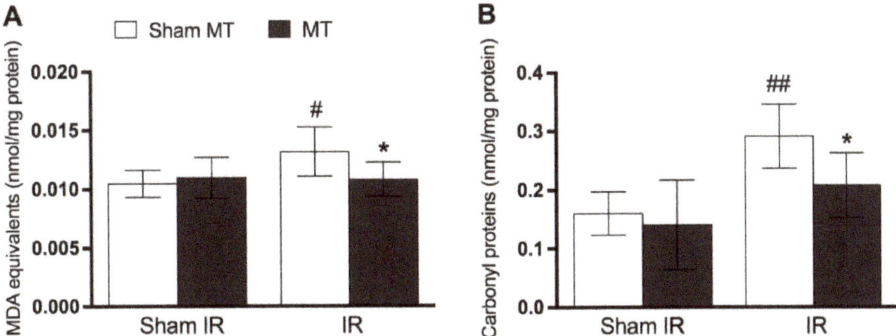

Figure 3. Evidence of the effects of MT on oxidative stress markers at the 2nd day after the IR procedure. Panels show the preventive effect of MT on the increase of MDA (panel **A**) and carbonyl proteins (panel **B**). Each point represents the mean of 8 animals and vertical lines show the SD. Statistical analyses were performed by one-way ANOVA followed by Newman–Keuls Multiple Comparison Test. The symbols denote a significant difference of * $p < 0.05$ when compared to IR + Sham MT group, # $p < 0.05$ or ## $p < 0.001$ when compared to Sham + Sham MT group. MT: Manual therapy, IR: Ischemia and reperfusion, MDA: Malondialdehyde.

Figure 4 shows that IR injury diminishes the levels of SOD (Figure 4A, $p = 0.03$) and CAT (Figure 4B, $p = 0.02$) activity in the animals' paw tissue on day 2 following IR. MT treatment effectively prevented the decrease in the activity of CAT (Figure 4B, $p = 0.02$), but not SOD (Figure 4A, $p = 0.31$) induced by IR.

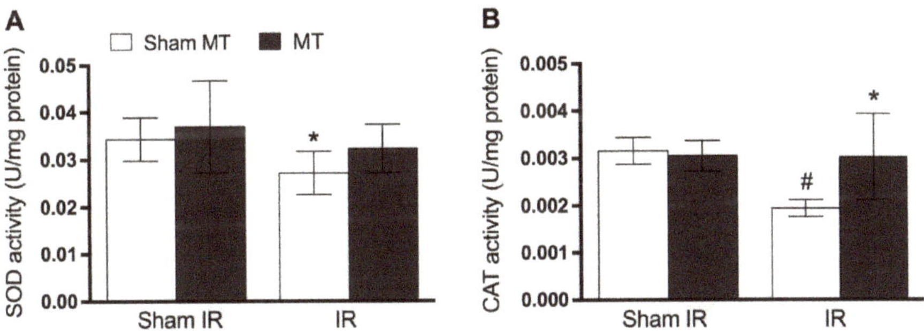

Figure 4. Evidence of MT effects on anti-oxidant enzymes levels at the 2nd day after IR procedure. Panel **A** shows that there was no significant difference on superoxide dismutase (SOD) activity, while panel **B** shows a significant difference on CAT activity. Each point represents the mean of 8 animals and vertical lines show the SD. Statistical analyses were performed by one-way ANOVA followed by Newman–Keuls Multiple Comparison Test. The symbols denote a significant difference of * $p < 0.05$ when compared to IR + Sham MT group, # $p < 0.05$ when compared to Sham + Sham MT group. MT: Manual therapy, IR: Ischemia and reperfusion, SOD: Superoxide dismutase, CAT: Catalase.

Figure 5 shows that IR injury diminishes Complex I ($p = 0.001$), II ($p = 0.001$) and IV activity ($p = 0.001$) in the animal hindpaw tissue on day 2 following IR. However, MT treatment was ineffective in preventing decreases in mitochondrial complex activity.

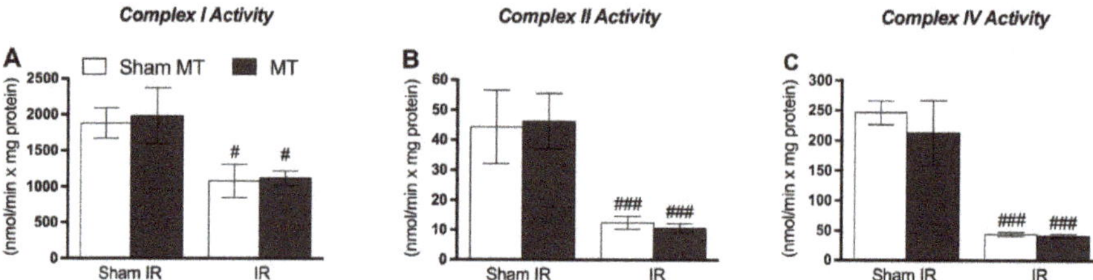

Figure 5. Effects of MT on mitochondrial function at the 2nd day after IR procedure. Complex I activity (**A**), Complex II activity (**B**) and complex IV activity (**C**) all panels show that IR reduces mitochondrial function but ankle joint mobilization could not prevent the mitochondrial function reduction. Each point represents the mean of 8 animals and vertical lines show the SD. Statistical analyses were performed by One-way ANOVA followed by Newman–Keuls multiple comparison test. The symbols denote a significant difference of $\#\ p < 0.05$ or $\#\#\#\ p < 0.001$ when compared to Sham + Sham MT group. MT: Manual therapy, IR: Ischemia and reperfusion.

4. Discussion

The results of this study show that MT produces analgesic and anti-oxidative effects in a murine model of CRPS-I. Our findings further show that the MT reduced mechanical hyperalgesia in all days evaluated, prevented the increase of TBARS and protein carbonyls concentrations, and prevented the reduction of CAT activity, while not influencing SOD activity. No effects from MT were observed in mitochondrial complex activity. The effect of MT has been demonstrated in multiple nerve injury models, such as the sciatic nerve crush injury model [42], postoperative pain model, and plantar incision surgery model [2], in which ankle joint mobilization produced an analgesic effect. The present set of experiments demonstrated for the first time an analgesic effect of MT in a murine CRPS-I model.

While CRPS pathophysiology is not fully understood, oxidative events are thought to give rise to primary afferent nociceptor sensitization which contributes to central sensitization. It has been well documented that prolonged hindlimb IR produces a subsequent cascade of inflammatory events, with pivotal roles being played by reactive oxygen species [12,13]. Ischemia-reperfusion results in production of oxidants, superoxide, hydroxyl radicals hydrogen peroxide, among other ROS initiated by the enzymes NADPH oxidase [43,44] or xanthine oxidase [45,46]. Coderre et al. [1] demonstrated that free radical scavengers reduced CPIP symptoms thereby emphasizing the important role that oxidants play in the maintenance of neuropathic pain-like symptoms in CRPS-I models [47]. Thus, the observed anti-oxidative effects of MT may be associated with the analgesia induced by MT in this current neuropathic pain model.

We observed that the IR procedure that induced mechanical hyperalgesia was maintained up to the 11th day of evaluation and that acute MT treatment was able to reduce mechanical hyperalgesia for 1h after treatment. Repeated treatments did not show a cumulative effect, since after MT treatments an increase in duration of analgesia was not observed. The specific analgesic mechanisms underlying peripheral joint mobilization remain unclear, but activation of inhibitory neuroreceptors such as opioid, cannabinoid 1(CB_1) and 2 (CB_2) receptors are all thought to play a role [2,25,26].

Possible explanations for the MT-related findings in the current study are the effects of the neuroreceptors (opioid, cannabinoid and adenosine receptors) activated by the oxidative system. In this context, it has been shown that the injection of opioid peptides in rats decrease the stress-induced activation of lipid peroxidation in plasma and liver tissue as well as increase catalase activity [27]. Interestingly, it has been shown in human monocytes/macrophages that during inflammation the CB_1 receptor is highly expressed and that its activation directly modulates inflammatory activity by means of production of ROS [28]. Conversely, CB_2 receptor activation exerts an anti-inflammatory response, such as inhibition of chemotactic movement in response to monocyte chemoattractant protein-1 (MCP-1) [48]. Moreover, the activation of CB_2 receptor may generate inhibitory signaling and directly suppress the production of ROS stimulated by the activation of the receptor CB_1 [29].

Recently, our research group has shown that MT reduces post-operative pain in mice by activation of the CB_2, but not the CB_1 receptor. That led us to believe that in the current study, the observed oxidative stress reduction (MDA and protein carbonyls) is due to the inhibitory effect on the activation of the CB_2 receptor mediated by MT on the ROS production stimulated by the activation of the CB_1 receptor.

The adenosinergic system also has been shown to modulate oxidative stress especially on activation of the A_1 receptor. The antioxidant effect of an adenosine A_1 receptor agonist cyclopentyladenosine (CPA) was recently studied in a focal cerebral ischemia model. Changes in lipid peroxidation (LPO) processes in the brain and blood tissue were demonstrated following ischemic brain injury. Changes in the ratio between LPO and antioxidant protection were less pronounced after cyclopentyladenosine treatment [30]. Current thought is that signaling activated by adenosine and/or other receptors (such as opioid or bradykinin) converge on key targets like mitochondrial K_{ATP} channels or the mitochondrial permeability transition pore (MPTP) [49–52]. The MPTP may be inhibited through control of protein phosphorylation (together with effects of K_{ATP} opening), or by inhibition of cellular oxidative stress and subsequent MPTP thiol modification [49,52]. Moreover, it has been shown that oxidative stress is selectively modulated endogenously by the A_1 receptor in ischemic hearts [52].

In parallel of this literature, Martins et al. [2] verified that MT reduces mechanical hyperalgesia induced by plantar incision surgery and these effects were mediated by the activation of A_1 receptors activation. Therefore, we may consider the hypothesis that in the present study, endogenous adenosine might have been secreted during MT which mediated analgesia and oxidative stress reduction. We found that ankle joint mobilization significantly reduced oxidative damage in the hindpaw, potentially suggesting a novel analgesic mechanism of MT by increased CAT activity in a CRPS-I model. These findings corroborate the study of Kolberg et al. [24] in humans, where they also observed that joint manipulation increased CAT activity in erythrocytes showing an anti-oxidative effect of manual therapy. In contrast to our findings, they did not find changes in lipidic peroxidation concentrations. These discrepancies may be explained by differences in the analyzed tissues/cells and/or the particular models used. The results of the present study are important in the clinical setting, since MT (joint mobilization) is widely used in the rehabilitation protocols of patients with chronic pain. In this sense, our findings suggest that MT may be used to treat CRPS-1 in humans, since it has an anti-oxidant effect. In addition, these results support the need for future clinical trials that associate MT with anti-oxidant therapies for the effective treatment of CRPS-I.

Limitations

This study did not evaluate the effect of MT on oxidative stress at other (non-peripheral) pain modulation sites such as the spinal cord, brainstem or sensory cortex which would allow a broader understanding of the effects of MT on CRPS-I. This study also did not analyze the oxidative stress parameters in the blood of mice, which would be interesting to investigate in a clinical setting. Future studies are needed to improve our understanding regarding the association between oxidative stress and the antihyperalgesic effects caused by MT and to establish the precise neurobiological systems underlying this effect of MT on oxidative stress parameters.

5. Conclusions

In summary, current results extended previous findings and demonstrated that daily sessions of MT presented antihyperalgesic effects mediated, at least in part, through (1) prevention of TBARS and protein carbonyls increase in peripheral (hindpaw) tissue and, (2) improvement of the antioxidant defense system (increase of CAT, but not SOD activity). MT did not change the analyzed mitochondrial complex activity. Together, these new findings contribute to a better understanding of the neurobiological mechanisms responsible for the therapeutic effect of MT, as well as provide additional support for its use as adjunctive treatment of CRPS-I.

Author Contributions: All authors conceptualized this study; D.F.M. and L.M.-M. concept/idea/research design; A.S.I.S., L.A.C., D.D.L., D.C.S., W.R.R., D.F.M. writing; J.S., D.D.L., A.C.C.K. data collection; D.F.M., L.G.D., A.G., L.R.S., G.T.R., D.F., F.P. analyzed data; D.F.M. fund procurement; L.M.-M., A.R.S.S. and D.F.M. providing facilities/equipment; D.F.M., L.M.-M., A.S.I.S., W.R.R. consultation (including review of manuscript before submitting).

Funding: The present study was supported by grants from Coordenação de Aperfeiçoamento de Pessoal de Nível Superior (CAPES), Conselho Nacional de Desenvolvimento Científico e Tecnológico (CNPq—476454/2013-1 and CNPq- 309407/2017-6), and Fundação de Amparo à Pesquisa e Inovação do Estado de Santa Catarina (FAPESC–3414/2012 and FAPESC-2019TR73), Brazil.

Acknowledgments: We would like to thank to Universidade do Sul de Santa Catarina, Cursos de Fisioterapia e Medicina and Programa Unisul de Iniciação Científica (PUIC).

Conflicts of Interest: The authors declare no conflict of interest.

References

1. Coderre, T.J.; Xanthos, D.N.; Francis, L.; Bennett, G.J. Chronic post-ischemia pain (CPIP): A novel animal model of complex regional pain syndrome-Type I (CRPS-I; reflex sympathetic dystrophy) produced by prolonged hindpaw ischemia and reperfusion in the rat. *Pain* **2004**, *112*, 94–105. [CrossRef] [PubMed]

2. Martins, D.F.; Mazzardo-Martins, L.; Cidral-Filho, F.J.; Stramosk, J.; Santos, A.R.S. Ankle joint mobilization affects postoperative pain through peripheral and central adenosine A1 receptors. *Phys. Ther.* **2013**, *93*, 401–412. [CrossRef] [PubMed]

3. Lawand, N.B.; McNearney, T.; Westlund, K.N. Amino acid release into the knee joint: Key role in nociception and inflammation. *Pain* **2000**, *86*, 69–74. [CrossRef]

4. Lee, S.; Lee, J.; Kim, Y.; Choi, J.; Jeon, S.; Kim, D.; Jeong, H.; Lee, Y.; Park, H. Antiallodynic Effects of Bee Venom in an Animal Model of Complex Regional Pain Syndrome Type 1 (CRPS-I). *Toxins* **2017**, *9*, 285. [CrossRef] [PubMed]

5. Birklein, F.; Schmelz, M. Neuropeptides, neurogenic inflammation and complex regional pain syndrome (CRPS). *Neurosci. Lett.* **2008**, *437*, 199–202. [CrossRef] [PubMed]

6. Kingery, W.S. Role of neuropeptide, cytokine, and growth factor signaling in complex regional pain syndrome. *Pain Med.* **2010**, *11*, 1239–1250. [CrossRef] [PubMed]

7. Maihöfner, C.; Seifert, F.; Markovic, K. Complex regional pain syndromes: New pathophysiological concepts and therapies. *Eur. J. Neurol.* **2010**, *17*, 649–660. [CrossRef] [PubMed]

8. Enax-Krumova, E.K.; Lenz, M.; Frettlöh, J.; Höffken, O.; Reinersmann, A.; Schwarzer, A.; Westermann, A.; Tegenthoff, M.; Maier, C. Changes of the sensory abnormalities and cortical excitability in patients with complex regional pain syndrome of the upper extremity after 6 months of multimodal treatment. *Pain Med.* **2017**, *18*, 95–106. [CrossRef]

9. Goris, R.J.; Dongen, L.M.; Winters, H.A. Are toxic oxygen radicals involved in the pathogenesis of reflex sympathetic dystrophy? *Free Radic. Res. Commun.* **1987**, *3*, 13–18. [CrossRef]

10. Eisenberg, E.; Shtahl, S.; Geller, R.; Reznick, A.Z.; Sharf, O.; Ravbinovich, M.; Erenreich, A.; Nagler, R.M. Serum and salivary oxidative analysis in Complex Regional Pain Syndrome. *Pain* **2008**, *138*, 226–232. [CrossRef]

11. Millecamps, M.; Laferrire, A.; Ragavendran, J.V.; Stone, L.S.; Coderre, T.J. Role of peripheral endothelin receptors in an animal model of complex regional pain syndrome type 1 (CRPS-I). *Pain* **2010**, *151*, 174–183. [CrossRef] [PubMed]

12. Yassin, M.M.; D'sa, A.B.; Parks, T.G.; McCaigue, M.D.; Leggett, P.; Halliday, M.I.; Rowlands, B.J. Lower limb ischaemia-reperfusion injury alters gastrointestinal structure and function. *Br. J. Surg.* **1997**, *84*, 1425–1429. [CrossRef] [PubMed]

13. Blaisdell, F.W. The pathophysiology of skeletal muscle ischemia and the reperfusion syndrome: A review. *Cardiovasc. Surg.* **2002**, *10*, 620–630. [CrossRef]

14. Besse, J.L.; Gadeyne, S.; Galand-Desmé, S.; Lerat, J.L.; Moyen, B. Effect of vitamin C on prevention of complex regional pain syndrome type I in foot and ankle surgery. *Foot Ankle Surg.* **2009**, *15*, 179–182. [CrossRef] [PubMed]

15. Zollinger, P.E.; Tuinebreijer, W.E.; Breederveld, R.S.; Kreis, R.W. Can vitamin C prevent complex regional pain syndrome in patients with wrist fractures? A randomized, controlled, multicenter dose-response study. *J. Bone Jt. Surg. Am.* **2007**, *89*, 1424–1431. [CrossRef]

16. Oerlemans, H.M.; Goris, J.; De Boo, T.; Oostendorp, R. Do physical therapy and occupational therapy reduce the impairment percentage in reflex sympathetic dystrophy? *Am. J. Phys. Med. Rehab.* **1999**, *78*, 533–539. [CrossRef]

17. Smith, T.O. How effective is physiotherapy in the treatment of complex regional pain syndrome type I? A review of the literature. *Musculoskelet. Care* **2005**, *3*, 181–200. [CrossRef] [PubMed]

18. Van de Meent, H.; Oerlemans, M.; Bruggeman, A.; Klomp, F.; Van Dongen, R.; Frolke, J.P. Safety of "pain exposure" physical therapy in patients with complex regional pain syndrome type 1. *Pain* **2011**, *152*, 1431–1438. [CrossRef]

19. Sherry, D.D.; Wallace, C.A.; Kelley, C.; Kidder, M.; Sapp, L. Short- and long-term outcomes of children with complex regional pain syndrome type I treated with exercise therapy. *Clin. J. Pain* **1999**, *15*, 218–223. [CrossRef]

20. Moss, P.; Sluka, K.; Wright, A. The initial effects of knee joint mobilization on osteoarthritic hyperalgesia. *Man. Ther.* **2007**, *12*, 109–118. [CrossRef]

21. Bialosky, J.E.; Bishop, M.D.; Price, D.D.; Robinson, M.E.; George, S.Z. The Mechanisms of Manual Therapy in the Treatment of Musculoskeletal Pain: A Comprehensive Model. *Man. Ther.* **2010**, *14*, 531–538. [CrossRef]

22. Camarinos, J.; Marinko, L. Effectiveness of manual physical therapy for painful shoulder conditions: A systematic review. *J. Man. Manip. Ther.* **2009**, *17*, 206–215. [CrossRef] [PubMed]

23. Walston, Z.; Hernandez, L.; Yake, D. Utilization of manual therapy to the lumbar spine in conjunction with traditional conservative care for individuals with bilateral lower extremity complex regional pain syndrome: A case series. *Physiother. Theory Pract* **2018**, *6*, 1–8. [CrossRef]

24. Kolberg, C.; Horst, A.; Kolberg, A.; Belló-Klein, A.; Partata, W.A. Effects of High-Velocity, Low-Amplitude Manipulation on Catalase Activity in Men with Neck Pain. *J. Manip. Physiol. Ther.* **2010**, *33*, 300–307. [CrossRef]

25. Martins, D.F.; Bobinski, F.; Mazzardo-Martins, L.; Cidral-filho, F.J.; Nascimento, F.P.; Gadotti, V.M. Ankle Joint Mobilization Decreases Hypersensitivity by Activation of Peripheral Opioid Receptors in a Mouse Model of Postoperative Pain. *Pain Med.* **2012**, *13*, 1049–1058. [CrossRef]

26. Martins, D.F.; Mazzardo-Martins, L.; Cidral-Filho, F.J.; Gadotti, V.M.; Santos, A.R.S. peripheral and spinal activation of cannabinoid receptors by joint mobilization alleviates postoperative pain in mice. *Neurosc.* **2013**, *255*, 110–121. [CrossRef]

27. Solin, A.V.; Liashev, I.D. Opioid peptides effect on lipid peroxidation in long-term stress. *Ross. Fiziol. Zhurnal Sechenova* **2012**, *98*, 1016–1020.

28. Han, K.H.; Lim, S.; Ryu, J.; Lee, C.W.; Kim, Y.; Kang, J.H.; Kang, S.S.; Ahn, Y.K.; Park, C.S.; Kim, J.J. CB_1 and CB_2 cannabinoid receptors differentially regulate the production of reactive oxygen species by macrophages. *Cardiovasc. Res.* **2009**, *84*, 378–386. [CrossRef]

29. Raborn, E.S.; Marciano-Cabral, F.; Buckley, N.E.; Martin, B.R.; Cabral, G.A. The cannabinoid delta-9-tetrahydrocannabinol mediates inhibition of macrophage chemotaxis to RANTES/CCL5: Linkage to the CB_2 receptor. *J. NeuroImmune Pharmacol.* **2008**, *3*, 117–129. [CrossRef]

30. Sufianova, G.Z.; Sufianova, A.A.; Shapkin, A.G. Effect of cyclopentyladenosine on lipid peroxidation during focal cerebral ischemia. *Bull. Exp. Biol. Med.* **2014**, *157*, 228–230. [CrossRef]

31. Zimmermann, M. Ethical guidelines for investigations of experimental pain in conscious animals. *Pain* **1983**, *16*, 109–110. [CrossRef]

32. Martins, D.F.; Mazzardo-Martins, L.; Soldi, F.; Stramosk, J.; Piovezan, A.P.; Santos, A.R.S. High-intensity swimming exercise reduces neuropathic pain in an animal model of complex regional pain syndrome type I: Evidence for a role of the adenosinergic system. *Neuroscience* **2013**, *234*, 69–76. [CrossRef]

33. Nucci, C.; Mazzardo-Martins, L.; Stramosk, J.; Brethanha, L.C.; Pizzolatti, M.G.; Santos, A.R.; Martins, D.F. Oleaginous extract from the fruits Pterodon pubescens Benth induces antinociception in animal models of acute and chronic pain. *J. Ethnopharmacol.* **2012**, *143*, 170–178. [CrossRef]

34. Esterbauer, H.; Cheeseman, K.H. Determination of aldehydic lipid peroxidation products: Malonaldehyde and 4-hydroxynonenal. *Methods Enzymol.* **1990**, *186*, 407–421. [CrossRef]

35. Levine, R.L.; Garland, D.; Oliver, C.N.; Amici, A.; Climent, I.; Lenz, A.G.; Ahn, B.W.; Shaltiel, S.; Stadtman, E.R. Determination of carbonyl content in oxidatively modified proteins. *Methods Enzymol.* **1990**, *186*, 464–478. [CrossRef]

36. Aebi, H. Catalase in vitro. *Meth Enzymol.* **1984**, *105*, 121–126. [CrossRef]

37. Bannister, W.H.; Bannister, J.V. Factor Analysis of the Activities of Superoxide Dismutase, Catalase and Glutathione Peroxidase in Normal Tissues and Neoplastic Cell Lines. *Free Radic. Res. Commun.* **1987**, *4*, 1–13. [CrossRef]

38. Lowry, O.H.; Rosebrough, N.J.; Farr, A.L.; Randall, R.J. Protein measurement with the Folin phenol reagent. *J. Biol. Chem.* **1951**, *193*, 265–275.

39. Cassina, A.; Radi, R. Differential Inhibitory Action of Nitric Oxide and Peroxynitrite on Mitochondrial Electron Transport. *Arch. Biochem. Biophys.* **1996**, *328*, 309–316. [CrossRef]

40. Fischer, J.C.; Ruitenbeek, W.; Berden, J.A.; Trijbels, J.F.; Veerkamp, J.H.; Stadhouders, A.M.; Sengers, R.C.; Janssen, A.J. Differential investigation of the capacity of succinate oxidation in human skeletal muscle. *Clin. Chim. Acta* **1985**, *153*, 23–36. [CrossRef]

41. Rustin, P.; Chretien, D.; Bourgeron, T.; Gerard, B.; Rötig, A.; Saudubray, J.M.; Munnich, A. Biochemical and molecular investigations in respiratory chain deficiencies. *Clin. Chim. Acta* **1994**, *228*, 35–51. [CrossRef]

42. Martins, D.F.; Mazzardo-Martins, L.; Gadotti, V.M.; Nascimento, F.P.; Lima, D.A.; Speckhann, B.; Favretto, G.A.; Bobinski, F.; Cargnin-Ferreira, E.; Bressan, E.; et al. Ankle joint mobilization reduces axonotmesis-induced neuropathic pain and glial activation in the spinal cord and enhances nerve regeneration in rats. *Pain* **2011**, *152*, 2653–2661. [CrossRef]

43. Kellogg, E.W.; Fridovich, I. Superoxide, hydrogen peroxide, and singlet oxygen in lipid peroxidation by a xanthine oxidase system. *J. Biol. Chem.* **1975**, *250*, 8812–8817.

44. McCord, J.M. Oxygen-derived radicals: A link between reperfusion injury and inflammation. *Fed. Proc.* **1987**, *46*, 2402–2406.

45. Inauen, W.; Suzuki, M.; Granger, D.N. Mechanisms of cellular injury: Potential sources of oxygen free radicals in ischemia/reperfusion. *Microcirc. Endothel. Lymphat.* **1989**, *5*, 143–155.

46. Partrick, D.A.; Moore, F.A.; Moore, E.E.; Barnett, C.C.; Silliman, C.C. Neutrophil priming and activation in the pathogenesis of postinjury multiple organ failure. *New Horiz.* **1996**, *4*, 194–210.

47. Millecamps, M.; Coderre, T.J. Rats with chronic post-ischemia pain exhibit an analgesic sensitivity profile similar to human patients with complex regional pain syndrome—Type, I. *Eur. J. Pharmacol.* **2008**, *583*, 97–102. [CrossRef]

48. Montecucco, F.; Burger, F. CB$_2$ cannabinoid receptor agonist JWH-015 modulates human monocyte migration through defined intracellular signaling pathways. *Am. J. Physiol.-Heart Circ. Physiol.* **2008**, *294*, 1145–1155. [CrossRef]

49. Pan, H.L.; Wu, Z.Z.; Zhou, H.Y.; Chen, S.R.; Zhang, H.M.; Li, D.P. Modulation of pain transmission by G-protein-coupled receptors. *Pharmacol. Ther.* **2008**, *117*, 141–161. [CrossRef]

50. Boison, D. Adenosine as a neuromodulator in neurological diseases. *Curr. Opin. Pharmacol.* **2008**, *8*, 2–7. [CrossRef]

51. Aguiar, A.S., Jr.; Boemer, G.; Rial, D.; Cordova, F.M.; Mancini, G.; Walz, R.; De Bem, A.F.; Latini, A.; Leal, R.B.; Pinho, R.A.; et al. High-intensity physical exercise disrupts implicit memory in mice: Involvement of the striatal glutathione antioxidant system and intracellular signaling. *Neuroscience* **2010**, *171*, 1216–1227. [CrossRef]

52. Reichelt, M.E.; Shanu, A.; Willems, L.; Witting, P.K.; Ellis, N.A.; Blackburn, M.R.; Headrick, J.P. Endogenous adenosine selectively modulates oxidant stress via the A1 receptor in ischemic hearts. *Antioxid. Redox Signal.* **2009**, *11*, 2641–2650. [CrossRef]

Article

Upregulation of IL-1 Receptor Antagonist in a Mouse Model of Migraine

Salvo Danilo Lombardo [1], Emanuela Mazzon [2], Maria Sofia Basile [1], Eugenio Cavalli [2], Placido Bramanti [2], Riccardo Nania [1], Paolo Fagone [1], Ferdinando Nicoletti [1,*] and Maria Cristina Petralia [2]

1   Department of Biomedical and Biotechnological Sciences, University of Catania, Via S. Sofia 89, 95123 Catania, Italy
2   IRCCS Centro Neurolesi Bonino Pulejo, Strada Statale 113, C.da Casazza, 98124 Messina, Italy
*   Correspondence: ferdinic@unict.it; Tel.: +39-095-478-1270

Received: 9 July 2019; Accepted: 18 July 2019; Published: 19 July 2019

Abstract: Migraine is a disorder characterized by attacks of monolateral headaches, often accompanied by nausea, vomiting, and photophobia. Around 30% of patients also report aura symptoms. The cause of the aura is believed to be related to the cortical spreading depression (CSD), a wave of neuronal and glial depolarization originating in the occipital cortex, followed by temporary neuronal silencing. During a migraine attack, increased expression of inflammatory mediators, along with a decrease in the expression of anti-inflammatory genes, have been observed. The aim of this study was to evaluate the expression of inflammatory genes, in particular that of IL-1 receptor antagonist *(IL-1RN)*, following CSD in a mouse model of familial hemiplegic migraine type 1 (FHM-1). We show here that the expression of *IL-1RN* was upregulated after the CSD, suggesting a possible attempt to modulate the inflammatory response. This study allows researchers to better understand the development of the disease and aids in the search for new therapeutic strategies in migraine.

Keywords: migraine; IL-1RN; cortical spreading depression; mouse model

---

1. Introduction

Migraine is a common neurological disease, representing the fourth cause of years lived with disability (YLDs) for women, and the eighth for men [1–3]. It is characterized by attacks of monolateral headaches, associated with nausea as well as phono- and photophobia [4]. Around 30% of patients also report aura symptoms, which are mostly visual. The cause of the aura is believed to be related to the cortical spreading depression (CSD), a self-propagating wave of cellular depolarization from the occipital cortex, followed by a transitory neuronal silencing [5,6].

The manifestation of these attacks depends on a genetic predisposition, associated to environmental stimuli such as stress, hormones, meteorological changes, and sleep disorders. For this reason, it is difficult to identify the etiology and physiopathogenesis of migraine. Recent evidence also suggests that immunoinflammatory events may also play a role in the pathogenesis of migraine [7,8].

In particular, much attention has been given to the role of immune-system hormones, named cytokines, to the pathogenesis of migraine [9].

Based on preclinical and clinical studies, the cytokines have been divided into at least five subfamilies: the proinflammatory Th1/Th17 cytokines, the anti-inflammatory Th2/Th3 cytokines, and the Th9 cytokine, represented by IL-9 [10–12].

It has been shown that Th1 and Th17 cytokines primarily exert proinflammatory effects. They are produced by M1 macrophages, Th1 and Th17 cells, and include—among others—IL-1, TNF-alpha, IFN-gamma, IL-12, IL-18, IL-22, IL-23, and IL-17. They may contribute to the initiation of cell-mediated autoimmune diseases such as rheumatoid arthritis, type 1 diabetes, multiple sclerosis, and Guillain-Barré syndrome [10,11,13].

Anti-inflammatory cytokines (e.g., IL-4, IL-10, IL-13 IL-35, TGF-beta) are primarily produced from M2 macrophages and Th2 and Th3 cells. They decrease Th1- and Th17-mediated immunoinflammatory events, and are implicated in IgE-mediated allergic diseases and eosinophil-mediated pathologies [14–17].

Sometimes, such as in the case of systemic lupus erythematosus, it seems that the combined action of Th1/Th2 cytokines may be simultaneously involved in the pathogenesis of the disease [18]. The precise role of IL-9 in regulation of the immune responses is less well-defined, and is receiving a great deal of attention [12].

In addition, the function of cytokines is also finely regulated by endogenous antagonists, which have been described for most cytokines, such as soluble receptors, anti-cytokine autoantibody, and for IL-1, the IL-1 receptor antagonist [19–21]. These endogenous antagonists are usually produced during immune responses, and serve to control and downregulate excessive signaling of the cytokine through binding with its functionally active receptors expressed on the surface of the target cells. For example, we have shown that blood levels of IL-1ra are augmented during attacks of multiple sclerosis and are further augmented from the treatment of the patients with IFN-beta [22].

Cytokines have recently been associated to the etiology of migraine, even if conflicting results have been found [23–25]. The association between migraine and the interleukin-1 receptor antagonist variable number tandem repeat (IL-1RN VNTR) has previously been investigated, but no statistically significant differences were discovered [26].

A cross-talk between neurons and immune cells has also been reported, which may contribute to the generation of the pain [27]. Indeed, activated macrophages and other non-neuronal cells might induce a meningeal "sterile inflammation" [28,29], contributing to the pain symptoms [30,31]. Furthermore, it has been reported that inflammation in the trigeminal nerve territory is often observed during migraine attacks [32], so the acute administration of corticosteroids has been tested to block pain [33].

Familial hemiplegic migraine type 1 (FHM-1) is a monogenic type of migraine with aura caused by mutations in the *CACNA1A* gene which determine an alteration of the passage of Ca^{2+} ions in the cerebral cortex [34,35]. The FHM mouse model, generated by introducing the R192Q mutation into the endogenous *CACNA1A* gene [36], is used to study the physiopathology of FHM-1 [37,38]. R192Q KI mice have increased neuronal Ca^{2+} influx and augmented glutamate release [39], which may explain the increased susceptibility to CSD [36,39]. Data from animal models have shown that CSD is able to activate meningeal trigeminovascular neurons, generating the sensation of pain [40–42].

The aim of the present study was to determine the expression of inflammatory genes and in particular that of IL-1 receptor antagonist (*IL-1RN*) upon CSD, in a murine model of FHM1.

2. Materials and Methods

2.1. Dataset Selection and Analysis

In order to evaluate the brain expression levels of *IL-1RN* following CSD, we interrogated the GSE67933 dataset, obtained from the GEO dataset [43]. The dataset included whole-genome transcriptomic data from wild-type (WT) mice and transgenic mice carrying the CACNA1A R192Q missense mutation (FHM1 R192Q mice). Gene expression profiles were obtained from cortical tissue of FHM1 R192Q and control mice, 24 h after experimentally induced CSD [44]. Briefly, CSD was induced by seven applications of a cotton pellet soaked in 300 mM KCl on the dura overlaying the occipital cortex, while in sham animals 300 mM NaCl was applied instead of KCl [44]. Deep SAGE sequencing was used to generate the expression profiles, and the data were normalized using the trimmed mean of M-values (TMM) method.

Functionally correlated genes were obtained using the web-based software STRING [45] and visualized as a gene network. Relationships of genes in the network were defined in terms of co-expression, text-mining, biochemical data, curated pathway, and protein–protein interactions.

The confidence cut-off for showing interaction links was set as medium (0.4) and the maximum number of interactors in the first shell was set at 50.

Heatmapper [46] software was used to generate the expression heatmap of the functional related genes to *IL-1RN* and to perform hierarchical clustering of genes. Average linkage was used as clustering method and Euclidean distance as distance measurement.

Co-expression analysis was carried out using the CoExpress software [47], and gene enrichment for biological processes (BPs) and molecular functions (MFs) was performed using the web-based utility DAVID version 6.8 [48].

2.2. Statistical Analysis

Data are shown as mean ± standard deviation. Differences in *IL-1RN* expression among the experimental groups were evaluated by ANOVA (Student's *t*-test) with Bonferroni post-hoc test. Correlation analysis was performed using the Pearson's correlation test. Statistical analysis was performed using IBM SPSS Statistics 23 [49] and GraphPad Prism 8.0.2 [50].

3. Results

3.1. IL-1RN Expression after CSD

The GSE67933 dataset was used to determine the expression levels of *IL-1RN* following CSD in a murine model of FHM1 (Figure 1). No significant differences in *IL-1RN* levels were observed in FHM1 R192Q vs. wild-type (WT) cortex at basal condition (sham). No significant differences in *IL-1RN* expression were observed following CSD induction in WT mice as compared to sham-operated WT mice. Significantly higher levels of *IL-1RN* were observed upon CSD in FHM1 R192Q animals as compared to sham-operated mice ($p < 0.0001$) (Figure 1).

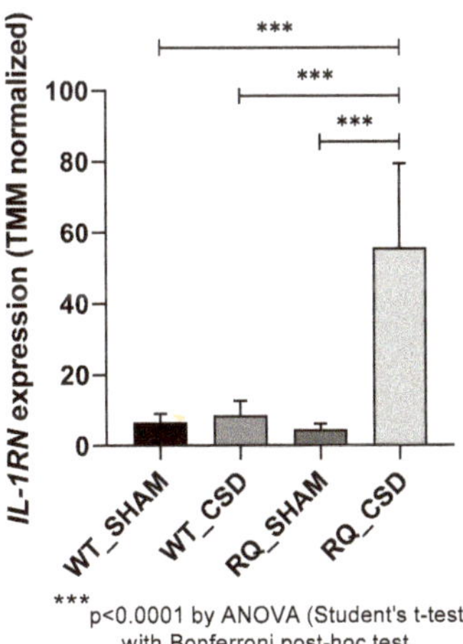

Figure 1. *IL-1RN* expression in a model of migraine. The expression of *IL-1RN* was investigated in mice bearing the R192Q mutation and wild–type (WT) mice at baseline and following CSD, as determined in the GSE67933 dataset. Data are presented as trimmed mean of M-values (TMM) normalized expression.

3.2. Identification of Genes Functionally Related to IL-1RN

We then evaluated the expression of the genes functionally related to *IL-1RN* (Figure 2A and Supplementary File 1). As shown in Figure 2 and Table 1, several of the genes identified were found to be modulated in the cortex from FHM1 R192Q mice subjected to CSD (Figure 2B). Among them, a significant upregulation of *IL-6*, *TNF*, *TLR2*, and *TLR4* could be observed in FHM1 R192Q mice upon CSD as compared to FHM1 R192Q sham-operated mice (Supplementary Table S1). Moreover, as compared to CSD-induced WT mice, FHM1 R192Q mice upon CSD expressed lower levels of *IL-18, IL-10, IL-2, IL-4,* and *IL-13,* and significantly higher levels of *IFNK, IL-17A, SOCS3,* and of several members of the chemokine/chemokine receptor family (*CCL2, CXCL2, CXCL10, CCL4, CCL7, CXCL1, CXCL3, CCR1, CXCL9*).

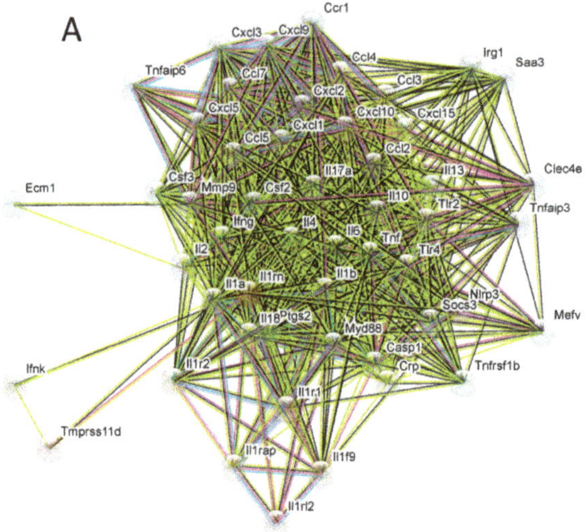

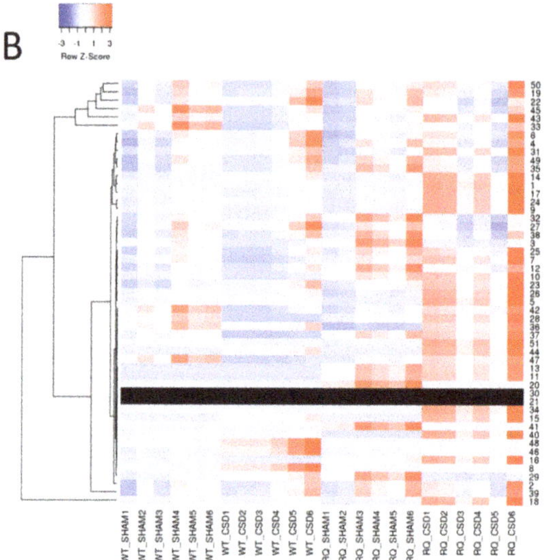

Figure 2. *IL-1RN* functionally-related genes. (**A**) Gene network of the *IL-1RN* functionally related genes. The edges connecting the nodes represent the interactions between genes, in terms of co-expression (black), text mining (light green), protein homology (cyan), association in curated database (light blue), and high-throughput experiments (purple). Empty nodes represent proteins of uncharacterized 3D structure, while filled nodes represent proteins with known or predicted tertiary structure. (**B**) Expression heatmap for the top correlated genes to *IL-1RN*, as determined in the GSE67933 dataset. Average linkage was used as clustering method and Euclidean distance as distance measurement. Tree branches represent the distance between genes.

3.3. Identification of Genes Statistically Correlated to IL-1RN

We found 927 genes statistically correlated to *IL-1RN* in CDS-induced FHM1 R192Q mice, considering as threshold a Pearson correlation coefficient (r) > 0.95 and a *p*-value < 0.05. The top 20 statistically significant correlated genes are presented in Table 1. Gene Ontology for biological processes (BPs) revealed a significant enrichment of genes involved in the "G-protein coupled receptor signaling pathway" (*p* < 0.0001), "sensory perception of smell" (*p* < 0.0001) and "inflammatory response" (*p* < 0.0001) (Figure 3A). The top enriched molecular functions (MFs) were: "olfactory receptor activity" (*p* < 0.0001), G-protein coupled receptor activity (*p* < 0.0001) and "odorant binding" (*p* < 0.0001) (Figure 3B).

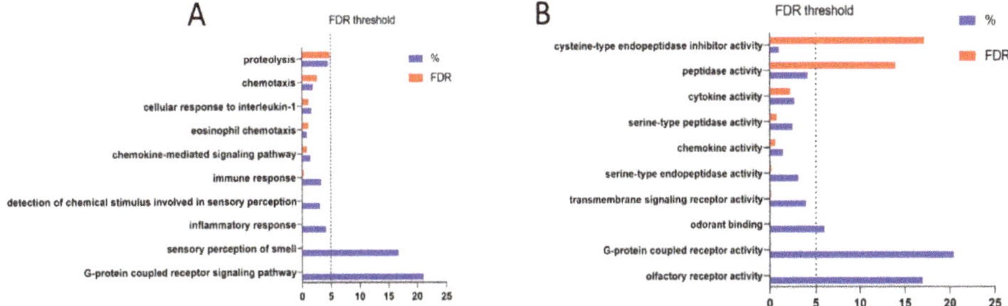

Figure 3. Gene ontology analysis for the genes statistically correlated to *IL-1RN* in the model of cortical spreading depression (CSD)-induced migraine. (**A**) Most enriched biological processes; (**B**) Most enriched molecular functions. FDR: False Discovery Rate

Table 1. Top 20 genes statistically correlated to *IL-1RN*.

Gene	Gene Stable ID	Pearson r	95% Confidence Interval	*p*-Value	R Squared
Cdhr5	ENSMUSG00000025497	0.9996	0.9991 to 0.9998	<0.0001	0.9992
Gm8251	ENSMUSG00000091844	0.9996	0.9991 to 0.9998	<0.0001	0.9993
Scarna3b	ENSMUSG00000088158	0.9993	0.9983 to 0.9997	<0.0001	0.9986
Nr0b1	ENSMUSG00000025056	0.9991	0.9980 to 0.9996	<0.0001	0.9983
Lrrc15	ENSMUSG00000052316	0.9987	0.9968 to 0.9994	<0.0001	0.9973
Ifi204	ENSMUSG00000073489	0.9982	0.9959 to 0.9993	<0.0001	0.9965
Cd200r3	ENSMUSG00000036172	0.9981	0.9955 to 0.9992	<0.0001	0.9962
Gm13389	ENSMUSG00000087079	0.9977	0.9947 to 0.9990	<0.0001	0.9955
Hpx	ENSMUSG00000030895	0.9976	0.9944 to 0.9990	<0.0001	0.9953
Gm15941	ENSMUSG00000086992	0.9974	0.9940 to 0.9989	<0.0001	0.9949
Zc3h12d	ENSMUSG00000039981	0.9973	0.9938 to 0.9989	<0.0001	0.9947
Gm49339	ENSMUSG00000062593	0.9973	0.9936 to 0.9988	<0.0001	0.9945
Klre1	ENSMUSG00000050241	0.9971	0.9933 to 0.9988	<0.0001	0.9943
Klk9	ENSMUSG00000047884	0.9966	0.9921 to 0.9986	<0.0001	0.9933
Gm22486	ENSMUSG00000080465	0.9966	0.9921 to 0.9986	<0.0001	0.9933
Abo	ENSMUSG00000015787	0.9965	0.9919 to 0.9985	<0.0001	0.9931
Cnga3	ENSMUSG00000026114	0.9965	0.9917 to 0.9985	<0.0001	0.9929
Ccl4	ENSMUSG00000018930	0.9960	0.9906 to 0.9983	<0.0001	0.9920
Snord66	ENSMUSG00000077239	0.9958	0.9901 to 0.9982	<0.0001	0.9916
Gm13429	ENSMUSG00000085141	0.9957	0.9898 to 0.9982	<0.0001	0.9913

4. Discussion

Many studies have shown a relationship between migraine and inflammation [51,52]. Neurogenic inflammation is characterized by the release of vasoactive neuropeptides from nociceptive sensory nerve terminals, including calcitonin gene-related peptide (CGRP), substance P (SP), and neurokinin A. These peptides lead to the dilatation of vessels, with increased permeability and consequent exudation of fluids, plasma proteins, leukocyte extravasation, and mast cell degranulation [51]. In particular, it has been proposed that migraine may be associated with an inflammation of the meninges, especially the dura mater. During a migraine attack, an idiopathic activation of the trigeminal sensory afferents is thought to facilitate the nociceptive transmission to the central nervous system (CNS). Accordingly, inhibition of the dural neurogenic inflammation with molecules able to inhibit the pathways involved in the activation and sensitization of trigeminovascular neurons—at both their central and peripheral perivascular nerve endings—have been tested as potential therapeutic strategies in the treatment of migraine [51].

The main current therapy for migraine is based on triptans [53], which have been shown to attenuate the release of neuropeptides and neurogenic plasma protein extravasation. These findings provide support for the validity of using animal models of neurogenic inflammation to investigate putative etiopathogenic mechanisms in migraine.

During the interictal period (headache-free days), independent studies have demonstrated increased peripheral levels of the pro-inflammatory cytokines IL-1β, IL-6, and TNF-α, and of the chemokine IL-8. On the other hand, the levels of the anti-inflammatory cytokine IL-10 have been found to be either unaltered or reduced in migraine patients as compared to healthy controls. Moreover, during a migraine attack, the serum concentrations of IL-1β, IL-6, IL-8, and TNF-α increase, alongside the levels of IL-10; in contrast, the type 2 cytokines IL-4 and IL-5 decrease [54]. Interestingly, migraine seems to be associated with immune-inflammatory and atopic diseases sustained by both Th1- or Th2-dominant responses, such as Inflammatory Bowel Disease and asthma [54].

Many studies have already observed that CSD induces the upregulation of several pro-inflammatory cytokines [6,40,44,55–58]. In the present study, we evaluated the expression of *IL-1RN* during CSD in a model of FMH1 by using a publicly available deep SAGE dataset. The use of whole-genome expression data has been extensively used for the identification of novel pathogenic pathways and therapeutic targets in a variety of diseases (e.g., autoimmunity [59–62], cancer [63–66], hepatic [67,68], neurodegenerative, and infectious diseases [69]).

The protein encoded by *IL-1RN* is a soluble factor that regulates the inflammatory response, as reported by various studies [21,22,70,71]. However, its role in migraine is still largely unexplored. In a previous study, significantly higher levels of *IL-1RN*, *TGF-β1*, and *MCP-1* in the cerebrospinal fluid of migraine patients as compared to controls were found [72].

Unlike *IL-2*, *IL-4*, *IL-10*, and *IL-13*, the expression of *IL-1RN* and *IL-6* was higher in the brains of FHM1 R192Q mutant mice than in WT mice, probably due to an alteration in Ca^{2+} ion channels. We can speculate that these molecules may exert a homeostatic role aimed at counteracting ongoing immunoinflammatory events. Even if the precise biological mechanisms by which IL-1RN production is increased is not known, it could be explained by an induction promoted by both IL-6 [73] and IFN-α [74]. Indeed, a role for the IFN pathway in migraine has been described [44]. We may speculate that compensatory mechanisms may be working constantly in migraine patients via the reciprocal regulation of pro- and anti-inflammatory factors, leading to Th1-dominant responses and the consequent effects associated with these cytokines. Moreover, we found an important increase in several members of the chemokine/chemokine receptor family in FHM1 R192Q, which could contribute to the vasodilatation and to the swelling mechanism underlying migraine and nasal congestion, which represents one of the most common symptoms [75].

The role of inflammation in migraine is further supported by the therapeutic effects of non-steroidal anti-inflammatory drugs (NSAIDs), currently recommended as the first-line medications for acute migraine attacks, as they improve both pain and breathing [76,77]. Recently, monoclonal antibodies targeting the CGRP pathway (i.e. fremanezumab, eptinezumab, and galcanezumab) have also been tested in both chronic and episodic migraine, and were shown to improve migraine-associated symptoms, quality of life, and disability, and to reduce monthly migraine days [78–81].

This study may set the basis for new therapeutic strategies for the treatment of migraine, such as anakinra [82]. Anakinra is a recombinant nonglycosylated analogue of the human IL-1RN which competitively blocks the binding of IL-1ß and IL-1α to the IL-1 receptors. Anakinra was approved by the Food and Drug Administration (FDA) in 2012 for chronic infantile neurological cutaneous and articular syndrome (CINCA) and in 2013 by the European Medicines Agency (EMA) for all subtypes of cryopyrin-associated periodic syndrome (CAPS). In nonhuman primates, anakinra has been shown to be able to cross the blood–brain barrier in a dose-dependent manner. Therefore, a direct anti-inflammatory action in the CNS is plausible [83]. Although only marginally related to migraine, a recent prospective, open-label, long-term study in 43 patients with severe cryopyrin-associated periodic syndromes (CAPS) demonstrated that anakinra treatment significantly decreased central nervous system inflammation and headaches in pediatric patients.

5. Conclusions

Little is known about the etiology and physiopathogenesis of migraine, but a large body of evidence shows a relationship with inflammation. In this paper, we show the gene expression of pro- and anti-inflammatory cytokines in a mouse migraine model. In particular, we focus on the expression of *IL-1RN*, which appears to be over-expressed after the CSD, suggesting a possible attempt to modulate the inflammatory response. This study may be the first one that allows a better understanding of the development of the disease and may aid in the search for new therapeutic strategies in migraine.

Supplementary Materials: The following are available online at http://www.mdpi.com/2076-3425/9/7/172/s1, File 1: Genes functionally related to *IL1RN*, Table S1: Multiple comparison.

Author Contributions: Conceptualization, S.D.L., E.M., F.N., P.F., M.C.P.; methodology, S.D.L., P.F., F.N.; formal analysis, S.D.L., P.F., R.N.; writing—original draft preparation, S.D.L., E.C. and M.S.B.; writing—review and editing, E.M., M.S.B., F.N., and M.C.P.; project administration, E.M., P.B., F.N.; funding acquisition, E.M., P.B., and F.N. All authors read and approved the final version of the manuscript.

Funding: This study was supported by current research funds 2018 of IRCCS Centro Neurolesi "Bonino Pulejo", Messina-Italy.

Conflicts of Interest: The authors declare no conflict of interest.

Sample Availability: Data are available from the Gene Expression Ombnibus Database.

References

1. Stovner, L.J.; Hagen, K.; Jensen, R.; Katsarava, Z.; Lipton, R.B.; Scher, A.I.; Steiner, T.J.; Zwart, J.A. The global burden of headache: A documentation of headache prevalence and disability worldwide. *Cephalalgia* **2007**, *27*, 193–210. [CrossRef] [PubMed]
2. Vetvik, K.G.; MacGregor, E.A. Sex differences in the epidemiology, clinical features, and pathophysiology of migraine. *Lancet Neurol.* **2017**, *16*, 76–87. [CrossRef]
3. Russell, M.B.; Rasmussen, B.K.; Thorvaldsen, P.; Olesen, J. Prevalence and sex-ratio of the subtypes of migraine. *Int. J. Epidemiol.* **1995**. [CrossRef] [PubMed]
4. Headache Classification Committee of the International Headache Society (IHS). The International Classification of Headache Disorders, 3rd edition (beta version). *Cephalalgia* **2013**. [CrossRef]
5. Lauritzen, M. Pathophysiology of the migraine aura: The spreading depression theory. *Brain* **1994**, *117 Pt 1*, 199–210. [CrossRef]
6. Cui, Y.; Kataoka, Y.; Watanabe, Y. Role of cortical spreading depression in the pathophysiology of migraine. *Neurosci. Bull.* **2014**, *30*, 812–822. [CrossRef] [PubMed]
7. Longoni, M.; Ferrarese, C. Inflammation and excitotoxicity: Role in migraine pathogenesis. *Neurol. Sci.* **2006**. [CrossRef]
8. Edvinsson, L.; Haanes, K.A.; Warfvinge, K. Does inflammation have a role in migraine? *Nat. Rev. Neurol.* **2019**. [CrossRef]
9. Cacabelos, R.; Torrellas, C.; Fernández-Novoa, L.; López-Muñoz, F. Histamine and Immune Biomarkers in CNS Disorders. *Mediat. Inflamm.* **2016**, *2016*, 1–10. [CrossRef]
10. Roeleveld, D.M.; Koenders, M.I. The role of the Th17 cytokines IL-17 and IL-22 in Rheumatoid Arthritis pathogenesis and developments in cytokine immunotherapy. *Cytokine* **2015**, *74*, 101–107. [CrossRef]
11. Zhang, H.-L.; Zheng, X.-Y.; Zhu, J. Th1/Th2/Th17/Treg cytokines in Guillain–Barré syndrome and experimental autoimmune neuritis. *Cytokine Growth Factor Rev.* **2013**, *24*, 443–453. [CrossRef] [PubMed]

12. Chakraborty, S.; Kubatzky, K.F.; Mitra, D.K. An Update on Interleukin-9: From Its Cellular Source and Signal Transduction to Its Role in Immunopathogenesis. *Int. J. Mol. Sci.* **2019**, *20*, 2113. [CrossRef] [PubMed]

13. Dujmovic, I.; Mangano, K.; Pekmezovic, T.; Quattrocchi, C.; Mesaros, S.; Stojsavljevic, N.; Nicoletti, F.; Drulovic, J. The analysis of IL-1 beta and its naturally occurring inhibitors in multiple sclerosis: The elevation of IL-1 receptor antagonist and IL-1 receptor type II after steroid therapy. *J. Neuroimmunol.* **2009**, *207*, 101–106. [CrossRef] [PubMed]

14. Ayakannu, R.; Abdullah, N.A.; Radhakrishnan, A.K.; Lechimi Raj, V.; Liam, C.K. Relationship between various cytokines implicated in asthma. *Hum. Immunol.* **2019**. [CrossRef]

15. Nicoletti, F.; Mancuso, G.; Cusumano, V.; Di Marco, R.; Zaccone, P.; Bendtzen, K.; Teti, G. Prevention of endotoxin-induced lethality in neonatal mice by interleukin-13. *Eur. J. Immunol.* **1997**, *27*, 1580–1583. [CrossRef] [PubMed]

16. Conway, T.F.; Hammer, L.; Furtado, S.; Mathiowitz, E.; Nicoletti, F.; Mangano, K.; Egilmez, N.K.; Auci, D.L. Oral Delivery of Particulate Transforming Growth Factor Beta 1 and All-Trans Retinoic Acid Reduces Gut Inflammation in Murine Models of Inflammatory Bowel Disease. *J. Crohn's Colitis* **2015**, *9*, 647–658. [CrossRef]

17. Nicoletti, F.; Di Marco, R.; Patti, F.; Reggio, E.; Nicoletti, A.; Zaccone, P.; Stivala, F.; Meroni, P.L.; Reggio, A. Blood levels of transforming growth factor-beta 1 (TGF-beta1) are elevated in both relapsing remitting and chronic progressive multiple sclerosis (MS) patients and are further augmented by treatment with interferon-beta 1b (IFN-beta1b). *Clin. Exp. Immunol.* **1998**, *113*, 96–99. [CrossRef]

18. Barcellini, W.; Rizzardi, G.P.; Borghi, M.O.; Nicoletti, F.; Fain, C.; Del Papa, N.; Meroni, P.L. In vitro type-1 and type-2 cytokine production in systemic lupus erythematosus: Lack of relationship with clinical disease activity. *Lupus* **1996**, *5*, 139–145. [CrossRef]

19. Fiotti, N.; Giansante, C.; Ponte, E.; Delbello, C.; Calabrese, S.; Zacchi, T.; Dobrina, A.; Guarnieri, G. Atherosclerosis and inflammation. Patterns of cytokine regulation in patients with peripheral arterial disease. *Atherosclerosis* **1999**, *145*, 51–60. [CrossRef]

20. Nielsen, C.H.; Bendtzen, K. Immunoregulation by Naturally Occurring and Disease-Associated Autoantibodies: Binding to cytokines and their role in regulation of T-cell responses. *Adv. Exp. Med. Biol.* **2012**, *750*, 116–132. [CrossRef]

21. Arend, W.P.; Malyak, M.; Guthridge, C.J.; Gabay, C. INTERLEUKIN-1 RECEPTOR ANTAGONIST: Role in Biology. *Annu. Rev. Immunol.* **2002**. [CrossRef]

22. Nicoletti, F.; Patti, F.; Di Marco, R.; Zaccone, P.; Nicoletti, A.; Meroni, P.L.; Reggio, A. Circulating serum levels of IL-1ra in patients with relapsing remitting multiple sclerosis are normal during remission phases but significantly increased either during exacerbations or in response to IFN-β treatment. *Cytokine* **1996**. [CrossRef]

23. Perini, F.; D'Andrea, G.; Galloni, E.; Pignatelli, F.; Billo, G.; Alba, S.; Bussone, G.; Toso, V. Plasma cytokine levels in migraineurs and controls. *Headache* **2005**. [CrossRef] [PubMed]

24. Empl, M.; Sostak, P.; Riedel, M.; Schwarz, M.; Müller, N.; Förderreuther, S.; Straube, A. Decreased sTNF-RI in migraine patients? *Cephalalgia* **2003**. [CrossRef]

25. Covelli, V.; Munno, I.; Pellegrino, N.M.; Di Venere, A.; Jirillo, E.; Buscaino, G.A. Exaggerated spontaneous release of tumor necrosis factor-alpha/cachectin in patients with migraine without aura. *Acta Neurol. (Napoli)* **1990**, *12*, 257–263.

26. Yilmaz, I.A.; Özge, A.; Erdal, M.E.; Edgünlü, T.G.; Çakmak, S.E.; Yalin, O.Ö. Cytokine polymorphism in patients with migraine: Some suggestive clues of migraine and inflammation. *Pain Med.* **2010**. [CrossRef] [PubMed]

27. Skaper, S.D. Nerve growth factor: A neuroimmune crosstalk mediator for all seasons. *Immunology* **2017**, *151*, 1–15. [CrossRef] [PubMed]

28. Moskowitz, M.A. Neurogenic inflammation in the pathophysiology and treatment of migraine. *Neurology* **1993**, *43* (Suppl. S3), S16–S20. [PubMed]

29. Waeber, C.; Moskowitz, M.A. Migraine as an inflammatory disorder. *Neurology* **2012**. [CrossRef]

30. Vecchia, D.; Pietrobon, D. Migraine: A disorder of brain excitatory-inhibitory balance? *Trends Neurosci.* **2012**, *35*, 507–520. [CrossRef]

31. Ceruti, S.; Villa, G.; Fumagalli, M.; Colombo, L.; Magni, G.; Zanardelli, M.; Fabbretti, E.; Verderio, C.; van den Maagdenberg, A.M.J.M.; Nistri, A.; et al. Calcitonin Gene-Related Peptide-Mediated Enhancement of Purinergic Neuron/Glia Communication by the Algogenic Factor Bradykinin in Mouse Trigeminal Ganglia from Wild-Type and R192Q Cav2.1 Knock-In Mice: Implications for Basic Mechanisms of Migraine Pain. *J. Neurosci.* **2011**. [CrossRef]

32. Friedman, M.H. Local inflammation as a mediator of migraine and tension-type headache. *Headache* **2004**. [CrossRef]

33. Rowe, B.H.; Colman, I.; Edmonds, M.L.; Blitz, S.; Walker, A.; Wiens, S. Randomized controlled trial of intravenous dexamethasone to prevent relapse in acute migraine headache. *Headache* **2008**. [CrossRef]

34. Van Den Maagdenberg, A.M.J.M.; Haan, J.; Terwindt, G.M.; Ferrari, M.D. Migraine: Gene mutations and functional consequences. *Curr. Opin. Neurol.* **2007**, *20*, 299–305. [CrossRef]

35. Ophoff, R.A.; Terwindt, G.M.; Vergouwe, M.N.; Van Eijk, R.; Oefner, P.J.; Hoffman, S.M.G.; Lamerdin, J.E.; Mohrenweiser, H.W.; Bulman, D.E.; Ferrari, M.; et al. Familial hemiplegic migraine and episodic ataxia type-2 are caused by mutations in the Ca2+ channel gene CACNL1A4. *Cell* **1996**. [CrossRef]

36. Van Den Maagdenberg, A.M.J.M.; Pietrobon, D.; Pizzorusso, T.; Kaja, S.; Broos, L.A.M.; Cesetti, T.; Van De Ven, R.C.G.; Tottene, A.; Van Der Kaa, J.; Plomp, J.J.; et al. A Cacna1a knockin migraine mouse model with increased susceptibility to cortical spreading depression. *Neuron* **2004**. [CrossRef]

37. Jurkat-Rott, K.; Lerche, H.; Weber, Y.; Lehmann-Horn, F. Hereditary channelopathies in neurology. *Adv. Exp. Med. Biol.* **2010**, *686*, 305–334.

38. Tfelt-Hansen, P.C.; Koehler, P.J. One hundred years of migraine research: Major clinical and scientific observations from 1910 to 2010. *Headache* **2011**, *51*, 752–778. [CrossRef]

39. Tottene, A.; Conti, R.; Fabbro, A.; Vecchia, D.; Shapovalova, M.; Santello, M.; van den Maagdenberg, A.M.J.M.; Ferrari, M.D.; Pietrobon, D. Enhanced Excitatory Transmission at Cortical Synapses as the Basis for Facilitated Spreading Depression in CaV2.1 Knockin Migraine Mice. *Neuron* **2009**. [CrossRef]

40. Karatas, H.; Erdener, S.E.; Gursoy-Ozdemir, Y.; Lule, S.; Eren-Koçak, E.; Sen, Z.D.; Dalkara, T. Spreading depression triggers headache by activating neuronal Panx1 channels. *Science* **2013**, *339*, 1092–1095. [CrossRef]

41. Noseda, R.; Burstein, R. Migraine pathophysiology: Anatomy of the trigeminovascular pathway and associated neurological symptoms, cortical spreading depression, sensitization, and modulation of pain. *Pain* **2013**, *154* (Suppl. S1), S44–S53. [CrossRef]

42. Zhang, X.; Levy, D.; Kainz, V.; Noseda, R.; Jakubowski, M.; Burstein, R. Activation of central trigeminovascular neurons by cortical spreading depression. *Ann. Neurol.* **2011**. [CrossRef]

43. GEO DataSets. Available online: https://www.ncbi.nlm.nih.gov/gds (accessed on 15 April 2019).

44. Eising, E.; Shyti, R.; 't Hoen, P.A.C.; Vijfhuizen, L.S.; Huisman, S.M.H.; Broos, L.A.M.; Mahfouz, A.; Reinders, M.J.T.; Ferrari, M.D.; Tolner, E.A.; et al. Cortical Spreading Depression Causes Unique Dysregulation of Inflammatory Pathways in a Transgenic Mouse Model of Migraine. *Mol. Neurobiol.* **2017**. [CrossRef]

45. STRING. Available online: https://string-db.org/ (accessed on 15 April 2019).

46. Heatmapper. Available online: http://www.heatmapper.ca/ (accessed on 15 April 2019).

47. CoExpress software. Available online: http://www.bioinformatics.lu/coexpress (accessed on 15 April 2019).

48. DAVID version 6.8. Available online: http://david.abcc.ncifcrf.gov/ (accessed on 15 April 2019).

49. IBM SPSS Statistics 23. Available online: https://www.spss.it/ (accessed on 15 April 2019).

50. GraphPad Prism 8.0.2. Available online: https://www.graphpad.com/ (accessed on 15 April 2019).

51. Malhotra, R. Understanding migraine: Potential role of neurogenic inflammation. *Ann. Indian Acad. Neurol.* **2016**. [CrossRef]

52. Ramachandran, R. Neurogenic inflammation and its role in migraine. *Semin. Immunopathol.* **2018**, *40*, 301–314. [CrossRef]

53. Johnston, M.M.; Rapoport, A.M. Triptans for the management of migraine. *Drugs* **2010**, *70*, 1505–1518. [CrossRef]

54. Oliveira, A.B.; Bachi, A.L.L.; Ribeiro, R.T.; Mello, M.T.; Tufik, S.; Peres, M.F.P. Unbalanced plasma TNF-α and IL-12/IL-10 profile in women with migraine is associated with psychological and physiological outcomes. *J. Neuroimmunol.* **2017**, *313*, 138–144. [CrossRef]

55. Costa, C.; Tozzi, A.; Rainero, I.; Cupini, L.M.; Calabresi, P.; Ayata, C.; Sarchielli, P. Cortical spreading depression as a target for anti-migraine agents. *J. Headache Pain* **2013**. [CrossRef]

56. Thompson, C.S.; Hakim, A.M. Cortical Spreading Depression Modifies Components of the Inflammatory Cascade. *Mol. Neurobiol.* **2005**. [CrossRef]

57. Choudhuri, R.; Cui, L.; Yong, C.; Bowyer, S.; Klein, R.M.; Welch, K.M.A.; Berman, N.E.J. Cortical spreading depression and gene regulation: Relevance to migraine. *Ann. Neurol.* **2002**. [CrossRef]

58. Urbach, A.; Bruehl, C.; Witte, O.W. Microarray-based long-term detection of genes differentially expressed after cortical spreading depression. *Eur. J. Neurosci.* **2006**. [CrossRef]

59. Fagone, P.; Mazzon, E.; Cavalli, E.; Bramanti, A.; Petralia, M.C.; Mangano, K.; Al-Abed, Y.; Bramati, P.; Nicoletti, F. Contribution of the macrophage migration inhibitory factor superfamily of cytokines in the pathogenesis of preclinical and human multiple sclerosis: In silico and in vivo evidences. *J. Neuroimmunol.* **2018**, *322*. [CrossRef]

60. Fagone, P.; Mazzon, E.; Mammana, S.; Di Marco, R.; Spinasanta, F.; Basile, M.S.; Petralia, M.C.; Bramanti, P.; Nicoletti, F.; Mangano, K. Identification of CD4+ T cell biomarkers for predicting the response of patients with relapsing-remitting multiple sclerosis to natalizumab treatment. *Mol. Med. Rep.* **2019**, *20*, 678–684. [CrossRef]

61. Mammana, S.; Bramanti, P.; Mazzon, E.; Cavalli, E.; Basile, M.S.; Fagone, P.; Petralia, M.C.; McCubrey, J.A.; Nicoletti, F.; Mangano, K. Preclinical evaluation of the PI3K/Akt/mTOR pathway in animal models of multiple sclerosis. *Oncotarget* **2018**, *9*, 8263–8277. [CrossRef]

62. Fagone, P.; Muthumani, K.; Mangano, K.; Magro, G.; Meroni, P.L.; Kim, J.J.; Sardesai, N.Y.; Weiner, D.B.; Nicoletti, F. VGX-1027 modulates genes involved in lipopolysaccharide-induced Toll-like receptor 4 activation and in a murine model of systemic lupus erythematosus. *Immunology* **2014**, *142*. [CrossRef]

63. Presti, M.; Mazzon, E.; Basile, M.S.; Petralia, M.C.; Bramanti, A.; Colletti, G.; Bramanti, P.; Nicoletti, F.; Fagone, P. Overexpression of macrophage migration inhibitory factor and functionally-related genes, D-DT, CD74, CD44, CXCR2 and CXCR4, in glioblastoma. *Oncol. Lett.* **2018**, *16*, 2881–2886. [CrossRef]

64. Fagone, P.; Caltabiano, R.; Russo, A.; Lupo, G.; Anfuso, C.D.; Basile, M.S.; Longo, A.; Nicoletti, F.; De Pasquale, R.; Libra, M.; et al. Identification of novel chemotherapeutic strategies for metastatic uveal melanoma. *Sci. Rep.* **2017**, *7*, 44564. [CrossRef]

65. Basile, M.S.; Mazzon, E.; Russo, A.; Mammana, S.; Longo, A.; Bonfiglio, V.; Fallico, M.; Caltabiano, R.; Fagone, P.; Nicoletti, F.; et al. Differential modulation and prognostic values of immune-escape genes in uveal melanoma. *PLoS ONE* **2019**, *14*, e0210276. [CrossRef]

66. Mangano, K.; Mazzon, E.; Basile, M.S.; Di Marco, R.; Bramanti, P.; Mammana, S.; Petralia, M.C.; Fagone, P.; Nicoletti, F. Pathogenic role for macrophage migration inhibitory factor in glioblastoma and its targeting with specific inhibitors as novel tailored therapeutic approach. *Oncotarget* **2018**, *9*, 17951–17970. [CrossRef]

67. Fagone, P.; Mangano, K.; Mammana, S.; Pesce, A.; Pesce, A.; Caltabiano, R.; Giorlandino, A.; Portale, T.R.; Cavalli, E.; Lombardo, G.A.G.; et al. Identification of novel targets for the diagnosis and treatment of liver fibrosis. *Int. J. Mol. Med.* **2015**, *36*, 747–752. [CrossRef]

68. Mangano, K.; Cavalli, E.; Mammana, S.; Basile, M.S.; Caltabiano, R.; Pesce, A.; Puleo, S.; Atanasov, A.G.; Magro, G.; Nicoletti, F.; et al. Involvement of the Nrf2/HO-1/CO axis and therapeutic intervention with the CO-releasing molecule CORM-A1, in a murine model of autoimmune hepatitis. *J. Cell. Physiol.* **2018**, *233*, 4156–4165. [CrossRef]

69. Fagone, P.; Nunnari, G.; Lazzara, F.; Longo, A.; Cambria, D.; Distefano, G.; Palumbo, M.; Nicoletti, F.; Malaguarnera, L.; Di Rosa, M. Induction of OAS gene family in HIV monocyte infected patients with high and low viral load. *Antivir. Res.* **2016**, *131*. [CrossRef]

70. Arend, W.P.; Gabay, C. Physiologic role of interleukin-1 receptor antagonist. *Arthritis Res.* **2000**, *2*, 245–248. [CrossRef]

71. Freeman, B.D.; Buchman, T.G. Interleukin-1 receptor antagonist as therapy for inflammatory disorders. *Expert Opin. Biol.* **2005**. [CrossRef]

72. Bø, S.H.; Davidsen, E.M.; Gulbrandsen, P.; Dietrichs, E.; Bovim, G.; Stovner, L.J.; White, L.R. Cerebrospinal fluid cytokine levels in migraine, tension-type headache and cervicogenic headache. *Cephalalgia* **2009**. [CrossRef]

73. Tilg, H.; Trehu, E.; Atkins, M.B.; Dinarello, C.A.; Mier, J.W. Interleukin-6 (IL-6) as an anti-inflammatory cytokine: Induction of circulating IL-1 receptor antagonist and soluble tumor necrosis factor receptor p55. *Blood* **1994**, *83*, 113–118.

74.	Tilg, H.; Mier, J.W.; Vogel, W.; Aulitzky, W.E.; Wiedermann, C.J.; Vannier, E.; Huber, C.; Dinarello, C.A. Induction of circulating IL-1 receptor antagonist by IFN treatment. *J. Immunol.* **1993**, *150*, 4678–4692.

75.	Arslan, H.H.; Tokgöz, E.; Yildizolu, Ü.; Durmaz, A.; Bek, S.; Gerek, M. Evaluation of the changes in the nasal cavity during the migraine attack. *J. Craniofac. Surg.* **2014**. [CrossRef]

76.	Aukerman, G.; Knutson, D.; Miser, W.F. Management of the acute migraine headache. *Am. Fam. Physician* **2002**, *66*, 2123–2130.

77.	Hsu, Y.C.; Lin, K.C.; Wang, S.J.; Wang, P.J.; Wang, Y.F.; Lee, L.H.; Wu, Z.A.; Shih, C.S.; Chen, Y.Y.; Chen, W.H.; et al. Medical treatment guidelines for acute migraine attacks. *Acta Neurol. Taiwan.* **2017**, *26*, 78–96.

78.	Raffaelli, B.; Mussetto, V.; Israel, H.; Neeb, L.; Reuter, U. Erenumab and galcanezumab in chronic migraine prevention: Effects after treatment termination. *J. Headache Pain* **2019**, *20*, 66. [CrossRef]

79.	Silberstein, S.D.; Stauffer, V.L.; Day, K.A.; Lipsius, S.; Wilson, M.-C. Galcanezumab in episodic migraine: Subgroup analyses of efficacy by high versus low frequency of migraine headaches in phase 3 studies (EVOLVE-1 & EVOLVE-2). *J. Headache Pain* **2019**, *20*, 75. [CrossRef]

80.	Dodick, D.W.; Lipton, R.B.; Silberstein, S.; Goadsby, P.J.; Biondi, D.; Hirman, J.; Cady, R.; Smith, J. Eptinezumab for prevention of chronic migraine: A randomized phase 2b clinical trial. *Cephalalgia* **2019**. [CrossRef]

81.	Silberstein, S.D.; Cohen, J.M.; Yeung, P.P. Fremanezumab for the preventive treatment of migraine. *Expert Opin. Biol.* **2019**, 1–9. [CrossRef]

82.	Neven, B.; Marvillet, I.; Terrada, C.; Ferster, A.; Boddaert, N.; Couloignier, V.; Pinto, G.; Pagnier, A.; Bodemer, C.; Tardieu, M.; et al. Long-term efficacy of the interleukin-1 receptor antagonist anakinra in ten patients with neonatal-onset multisystem inflammatory disease/chronic infantile neurologic, cutaneous, articular syndrome. *Arthritis Rheum.* **2010**. [CrossRef]

83.	Landmann, E.C.; Walker, U.A. Pharmacological treatment options for cryopyrin-associated periodic syndromes. *Expert Rev. Clin. Pharm.* **2017**, *10*, 855–864. [CrossRef]

Article

Applying a Sensing-Enabled System for Ensuring Safe Anterior Cingulate Deep Brain Stimulation for Pain

Yongzhi Huang [1], Binith Cheeran [1], Alexander L. Green [1], Timothy J. Denison [2,*] and Tipu Z. Aziz [1,*]

[1]   Oxford Functional Neurosurgery Group, Nuffield Departments of Surgical Sciences, University of Oxford, Oxford OX3 9DU, UK
[2]   Institute of Biomedical Engineering, University of Oxford, Oxford OX3 7DQ, UK
*   Correspondence: timothy.denison@eng.ox.ac.uk (T.J.D.); tipu.aziz@nds.ox.ac.uk (T.Z.A.);
    Tel.: +44-1865617707 (T.J.D.); +44-1865227645 (T.Z.A.)

Received: 22 May 2019; Accepted: 25 June 2019; Published: 26 June 2019

Abstract: Deep brain stimulation (DBS) of the anterior cingulate cortex (ACC) was offered to chronic pain patients who had exhausted medical and surgical options. However, several patients developed recurrent seizures. This work was conducted to assess the effect of ACC stimulation on the brain activity and to guide safe DBS programming. A sensing-enabled neurostimulator (Activa PC + S) allowing wireless recording through the stimulating electrodes was chronically implanted in three patients. Stimulation patterns with different amplitude levels and variable ramping rates were tested to investigate whether these patterns could provide pain relief without triggering after-discharges (ADs) within local field potentials (LFPs) recorded in the ACC. In the absence of ramping, AD activity was detected following stimulation at amplitude levels below those used in chronic therapy. Adjustment of stimulus cycling patterns, by slowly ramping on/off (8-s ramp duration), was able to prevent ADs at higher amplitude levels while maintaining effective pain relief. The absence of AD activity confirmed from the implant was correlated with the absence of clinical seizures. We propose that AD activity in the ACC could be a biomarker for the likelihood of seizures in these patients, and the application of sensing-enabled techniques has the potential to advance safer brain stimulation therapies, especially in novel targets.

Keywords: deep brain stimulation; anterior cingulate cortex; seizures; after-discharge; local field potential; chronic pain

1. Introduction

Deep brain stimulation (DBS) is an established therapy for movement disorders. It has several advantages over lesioning techniques. For example, therapy-induced side effects are considered to be reversible, and DBS can be 'dosed' as symptoms evolve [1]. Interestingly, the first DBS surgeries were performed for chronic post-stroke pain [2]. Cingulotomy has historically been used to target the affective component of pain, for example intractable pain associated with terminal cancer [3–5]. However, side effects are common, such as impairments of attention and cognition [6,7]. As an alternative to destructive lesioning, anterior cingulate cortex (ACC) DBS implants were offered to patients with severe, medically-refractory pain, where established targets, such as sensory thalamus and periventricular/periaqueductal grey, had failed or where pain was too poorly localized to consider these targets [8,9].

The success of ACC stimulation in patients deemed refractory to other medical and surgical interventions was tempered by the onset of recurrent stereotyped neurological events after 12–60 months of active stimulation in approximately 18% of the patients [10]. Some patients experienced recurrent seizures, and stimulation had to be markedly reduced or switched off completely as a conservative solution to ameliorate side effects; at these settings, effective pain relief was lost [10]. However, how ACC stimulation affects brain function and induces seizures remains unclear.

Studies have suggested several neurophysiological biomarkers potentially associated with seizures, e.g., stimulation-elicited after-discharges (ADs) [11,12]. Cortical stimulation can induce ADs, sometimes followed by clinical seizures, whether or not those regions are known to cause spontaneous seizures [11,13,14]. Stimulation parameters for inducing ADs have shown considerable within- and between-subject variability, but in general ADs can be elicited with sufficient stimulus intensity and duration [15–17].

With the advent of recent implant technologies, the effects of DBS on brain neural activities can be chronically investigated by measurements of local field potentials (LFPs) in the brain. Prior animal studies have demonstrated the ability to detect AD activity in LFPs using implanted DBS leads in various brain regions [18,19]. Here, we applied this technology to investigate the effect of ACC stimulation on brain activity in patients with chronic pain and aimed to elucidate safe stimulation parameters that maintained adequate pain relief without inducing seizures.

2. Materials and Methods

2.1. Subjects

Three patients with chronic pain who experienced recurrent seizures during the course of DBS therapy were investigated. All patients initially underwent bilateral implantation of DBS electrodes (Model 3387, Medtronic®, Minneapolis, MN, USA) into the ACC at The John Radcliffe Hospital, Oxford, UK. The surgical procedure has been previously described [8]. Details of the patients are reported in Table 1. None of the patients had suffered seizures prior to their initial surgery. Pre-operative MRI and post-operative CT scans did not reveal any relevant structural abnormalities or complications, such as hemorrhage or ischemia.

However, stereotyped neurological events, clinically diagnosed as seizures, were reported in these patients after some period of effective therapy. Medical management with multiple anti-epileptics was only transiently effective (weeks to months) before seizures recurred. Patient 2 did not take anti-epileptic drugs as he preferred altering the stimulation settings rather than medication. Video-electroencephalograph telemetry (vEEG) was used to investigate and detect seizures, initially in the first case (Figure 1). Reducing stimulation amplitude to levels below the threshold for seizure induction, based on multiple vEEG tests and clinical review, eliminated both the clinical seizures and the benefit of pain relief. Despite the risk of seizure induction, the patients requested reinstatement of stimulation to re-capture pain relief.

Table 1. Demographics, etiologies, stimulation parameters, and clinical events of patients.

Patient	Age at Surgery/Sex	Etiology	Onset of Seizure after Surgery	Seizure Symptoms	DBS Settings at Onset of Seizures	Anti-Epileptic Drugs	DBS Settings with Seizure Free	Follow-Up
1	46/F	Whole spine pain secondary to multiple spinal interventions	20 months	1) Focal non-motor onset with impaired awareness 2) Nocturnal generalized tonic-clonic seizures (maximum frequency reported: 1 event per month, lasting up to 45 min)	5 V 130 Hz 450 µs	Levetiracetam Clobazam Sodium valproate	6 V 130 Hz 450 µs 8-s ramp 3 min ON/ 11 min OFF	Seizure free for 17 months
2	51/M	Whole body pain secondary to excision of ependymoma of cervical spinal cord	60 months	Focal non-motor onset with impaired awareness (maximum frequency reported: 1 event per h)	8.5 V 130 Hz 450 µs	no	6 V 130 Hz 450 µs 8-s ramp1 min ON/ 1 min OFF	Seizure free for 6 months
3	49/M	Right hemi body pain secondary to posterior fossa decompression for Arnold–Chiari malformation	12 months	Focal non-motor onset with impaired awareness (maximum frequency reported: 50 events per day)	8.5 V 130 Hz 450 µs	Levetiracetam Oxcarbazepine	6 V 130 Hz 450 µs 8-s ramp1 min ON/ 1 min OFF	Seizure free for 1 month (then system removed due to infection)

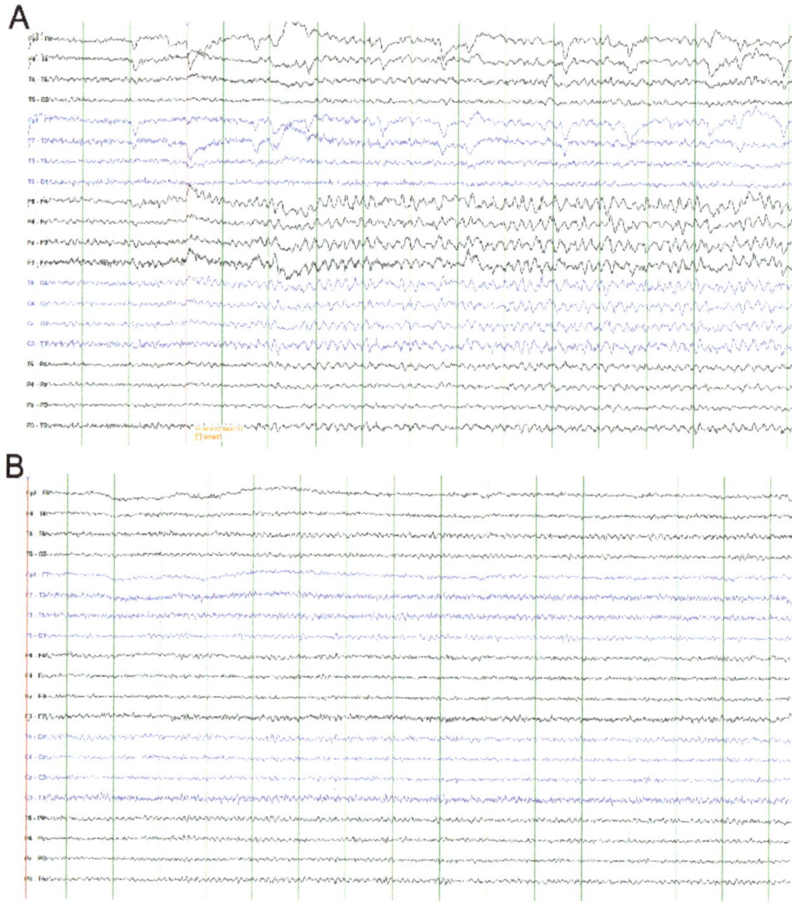

Figure 1. Example EEG recordings from patient 1. (**A**) Ictal EEG showing rhythmical, symmetrical 6 Hz theta slow wave activity across the frontocentral regions. (**B**) EEG showing normal background activity.

2.2. Stimulation and Local Field Potential Recordings

To further investigate the relationship between the stimulation and clinical events, a sensing-enabled neurostimulation system (Activa PC + S, Medtronic®) was implanted chronically, under the Medicines and Healthcare Products Regulatory Agency (MHRA) humanitarian exemption approval. This system allows for concurrent stimulation and recording from implanted DBS leads, and was used in a prior study to measure AD activity in an animal model, exploring network behavior in epilepsy [19].

Stimulation titration tests were systematically performed in patient 1 to investigate the effect of stimulation intensity on ACC neural activity. In particular, we explored whether ADs could be induced by ACC stimulation. Unilateral stimulation was increased from 0 V to target voltage (from 1 V to 6 V, 1 V steps) and then immediately switched off, with a stimulation-off interval of several seconds between steps. Ipsilateral and contralateral LFPs were recorded simultaneously during the same period. In addition, bilateral stimulation was also tested. The implantable pulse generator (IPG) was replaced with a second Activa PC + S system 13 months after the first implant of the Activa PC + S system due to depleted battery. Thereafter, bilateral stimulation at therapeutic amplitudes, using cycled stimulation with ramping, was explored to re-capture pain relief whilst minimizing seizures. Cycled stimulation on/off durations were selected based on results from the stimulation testing trials indicating that pain relief could be achieved with 3 min of stimulation, using therapeutic amplitudes, but could be lost 11 min later in this patient (Table 1). LFPs in the ACC were measured to investigate whether these stimulation patterns would induce ADs. Additional LFP recordings were collected during periods of chronic stimulation at home using the embedded loop recorder in the device [20,21].

The know-how learned in patient 1 was applied to patients 2 and 3. As unilateral ACC stimulation was found to be ineffective, bilateral stimulation without ramp, and with slowly ramped on/off, was tested, and LFPs were measured simultaneously. Subsequently, the stimulation pattern using cycling with ramp, shown through LFP measurements to avoid ADs, was applied for chronic treatment. Unfortunately, a month after IPG implant surgery, the sensing-enabled stimulation system had to be removed because of an infection in patient 3. During this period, the stimulation was clinically effective for pain relief without inducing seizures.

In all tests, electrical stimulation was delivered using a bipolar configuration between electrode contacts 0 (the deepest contact) and 3. Stimulation frequency and pulse width (typical therapy parameters 130 Hz, 450 µs) were fixed. All LFPs were recorded in a bipolar mode using the middle two contacts (1–2) of the electrodes (0.5 Hz pre-amplifier high-pass filtering, 100 Hz pre-amplifier low-pass filtering, 422 Hz sampling rate).

2.3. Data Analysis

LFP data were analyzed using custom scripts written in MATLAB (Version 9.1, MathWorks, Natick, MA, USA). To characterize the dynamic changes of neural activity, time-frequency representations of LFPs were performed using the short-time Fourier transform with a Hanning time window of 0.5 s and overlap of 0.45 s. These parameters provided a time resolution of 0.5 s and a frequency resolution of 2 Hz. Stimulation onset was identified as a period where observable high amplitude artifacts in the raw LFPs were accompanied by obvious 130 Hz stimulation frequency in the spectrograms. An AD episode was defined as the state with a sustained high amplitude, seizure-like activities in the raw LFPs and confirmed through elevated power across multiple frequency bands in the spectrograms.

Sensing channel saturation with large stimulation was a concern. To ensure a robust LFP measurement, a continuous monitoring approach was used to determine the reliability of the received signals [22]. Briefly, a continuous test tone at a discrete frequency (105 Hz) outside of the physiological band of interest was injected into the signal chain during recording through a parallel channel. If this tone's amplitude was compromised due to amplifier saturation, alternative signal chain parameters would be chosen, such as reducing the amplifier gain. For example, in our study, if the test tone shows amplifier saturation then the signal artifact following stimulation can look like a seizure activity (Figure 2A). In such cases, the amplifiers' gain was reduced to ensure the recording of reliable LFPs (Figure 2B).

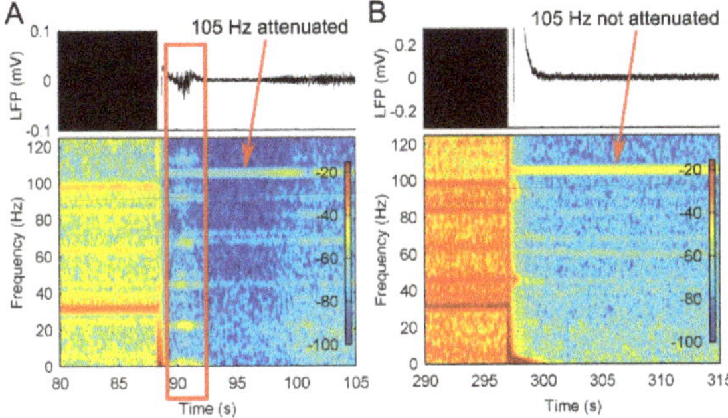

Figure 2. The use of a test tone to monitor the integrity of the bioelectric amplifier in measuring local field potentials (LFPs). (**A**) LFP measurement using an amplifier with a gain of 2000. The line box illustrates an example of distortion of signals induced by the amplifier recovering. (**B**) LFP measurement using an amplifier with a gain of 250.

3. Results

3.1. AD Activity in ACC LFPs Is Induced Following Stimulation

Stimulation titration tests revealed that the characteristics of LFP changes in the ACC were dependent upon the stimulation amplitudes. Figure 3 provides an illustration of the effects of unilateral stimulation amplitudes on ACC LFPs from patient 1. The ADs were induced following a stimulation at a threshold of 5 V on the left lead and 4 V on the right lead, respectively (Figure 3). Bilateral stimulation with 6 V amplitude on both leads also induced significant ADs in right ACC and slight ADs in left ACC (Figure 4A). The ADs in the ACC could also be observed when we repeated the titration tests after 1 month of bilateral therapeutic stimulation at 3.5 V (not shown). The AD threshold to stimulation was determined based on these tests. During the periods of measuring these LFPs, no clinical seizures were reported. Based on these LFP measurements, bilateral stimulation therapy with amplitudes below the AD threshold level was applied to attempt to prevent seizures and obtain pain relief. However, pain relief was inadequate, although no seizures were reported for approximately 12 months.

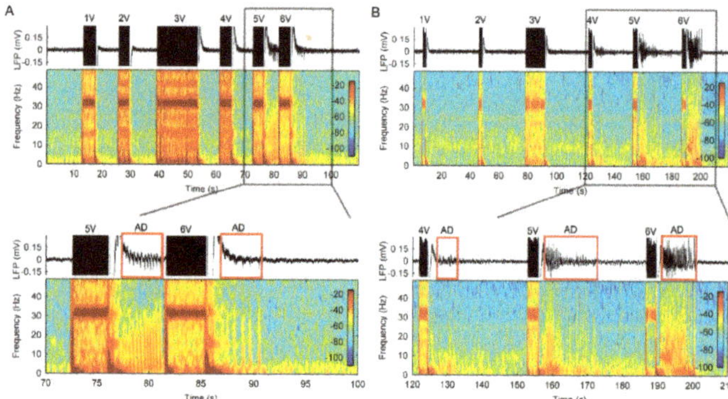

Figure 3. Effects of different unilateral stimulation amplitudes on the LFP recordings in the anterior cingulate cortex (ACC) from patient 1. (**A**) Bipolar LFP recordings and corresponding spectrograms from the left ACC during unilateral stimulation with increasing amplitudes. (**B**) Bipolar LFP recordings and corresponding spectrograms from the right ACC during unilateral stimulation with increasing amplitudes. Bottom panels show greater details illustrating several examples of after-discharges that occurred following cessation of stimulation.

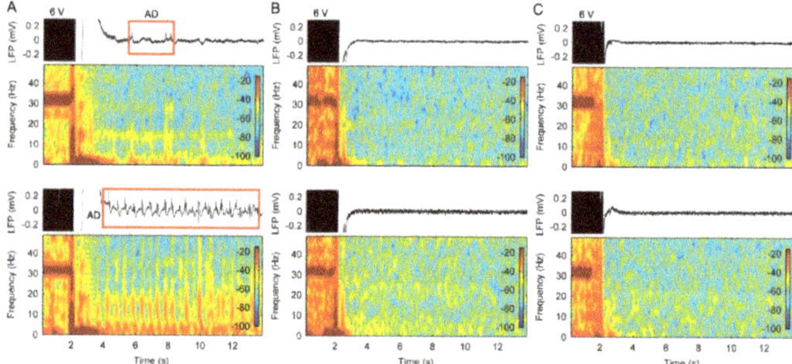

Figure 4. Effects of bilateral stimulation with/without a stimulus ramp on the LFP recordings in the bilateral ACC from patient 1. (**A**) Raw LFP recordings and corresponding spectrograms in the left (top panels) and right ACC (bottom panels) during stimulation without a stimulus ramp. (**B**) Raw LFP recordings and corresponding spectrograms in the left (top panels) and right ACC (bottom panels) during stimulation with a 4 s stimulus ramp. (**C**) Raw LFP recordings and corresponding spectrograms in the left (top panels) and right ACC (bottom panels) during stimulation with an 8 s stimulus ramp.

3.2. Stimulation with Slowly Ramped on/off during Cycling Successfully Eliminates ADs

Subsequently, based on the rationale detailed in the discussion, we tested a programming feature that allowed stimulation to be slowly ramped on/off during the cycling, rather than stimulation being started or stopped abruptly. When stimulation was delivered, using a pattern consisting of cycled stimulation, stimulation ramped down from the maximum amplitude to 0 V in 8 s and a high amplitude (6 V) that previously resulted in ADs, we did not observe the typical AD activities following the stimulation in patient 1 (Figure 4C). With these stimulation patterns for therapy, good pain relief was again reported, and the patient was discharged with the device programmed to collect additional LFP recordings over time.

LFPs recordings from patient 2 also showed that ADs were observed in the right ACC when using cycled stimulation patterns without a stimulus ramp and with a 4 s ramp; however, the AD activity disappeared when using stimulation with an 8 s ramp (Figure 5). The patient gained pain relief and no seizures during the test, using the cycled stimulation pattern at 6 V amplitude with an 8 s ramp. Therefore, we applied this stimulation pattern for chronic therapy in the patient.

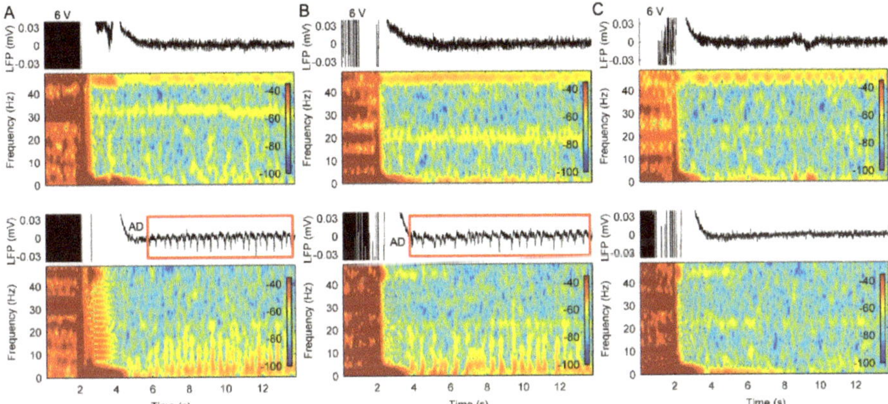

Figure 5. Effects of bilateral stimulation with/without a stimulus ramp on the LFP recordings in the bilateral ACC from patient 2. (**A**) Raw LFP recordings and corresponding spectrograms in the left (top panels) and right ACC (bottom panels) during stimulation without a stimulus ramp. (**B**) Raw LFP recordings and corresponding spectrograms in the left (top panels) and right ACC (bottom panels) during stimulation with a 4 s stimulus ramp. (**C**) Raw LFP recordings and corresponding spectrograms in the left (top panels) and right ACC (bottom panels) during stimulation with an 8 s stimulus ramp.

Although the system in patient 3 was removed due to infection at one month, the limited LFP recordings also revealed that there were no ADs using cycled stimulation with a ramp. Interestingly, in this case, we also did not observe the ADs when using stimulation without a ramp (Figure 6). Before system removal, the patient achieved pain relief and was seizure free.

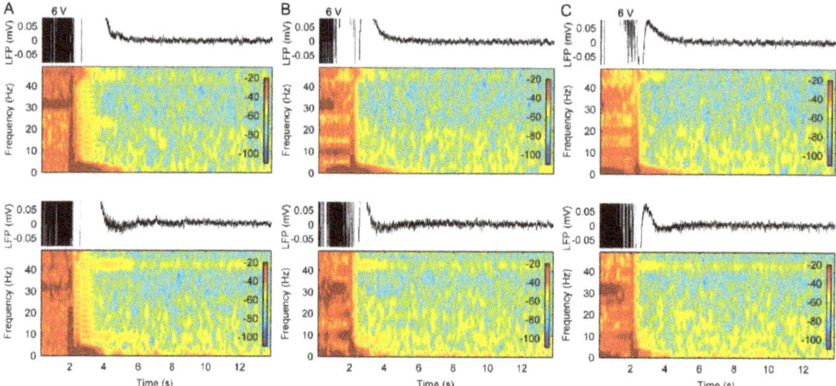

Figure 6. Effects of bilateral stimulation with/without a stimulus ramp on the LFP recordings in the bilateral ACC from patient 3. (**A**) Raw LFP recordings and corresponding spectrograms in the left (top panels) and right ACC (bottom panels) during stimulation without a stimulus ramp. (**B**) Raw LFP recordings and corresponding spectrograms in the left (top panels) and right ACC (bottom panels) during stimulation with a 4 s stimulus ramp. (**C**) Raw LFP recordings and corresponding spectrograms in the left (top panels) and right ACC (bottom panels) during stimulation with an 8 s stimulus ramp. Note that the after-effects observed during 3–6 s were confirmed to be due to sensing channel recovery using the test-tone method.

3.3. *The Use of Cycled Stimulation with Slow Ramps Provides Sustained Pain Relief without Seizures*

At the follow up, the parameter settings of stimulation that provided pain relief without ADs resulted in sustained therapeutic benefit without side-effects. The long-term LFP recordings, obtained during periods of chronic stimulation at home in patient 1, also showed no indication of ADs being triggered by chronic-cycled stimulation with a ramp (Figure 7). At the last follow-up, patient 1 had been seizure free (self-reported) for 17 months and patient 2 had been seizure free (self-reported) for 6 months.

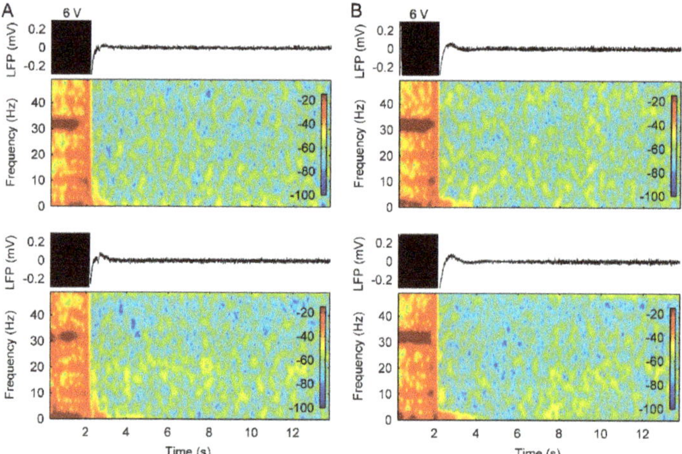

Figure 7. Long-term effects of bilateral therapeutic stimulation with an 8 s ramp on the LFP recordings in the bilateral ACC from patient 1. (**A**) Raw LFP recordings and corresponding spectrograms in the left (top panels) and right ACC (bottom panels) during stimulation for 5 months. (**B**) Raw LFP recordings and corresponding spectrograms in the left (top panels) and right ACC (bottom panels) during stimulation for 9 months.

4. Discussion

Neuromodulation can be an effective approach to pain management in patients that have exhausted medical therapies. However, the risk of adverse events with cortical stimulation, as reported here, needs to be addressed. For example, epidural motor cortex stimulation and repetitive transcranial magnetic stimulation have been explored for a variety of pain syndromes with variable success, but the induction of seizures has been reported as one of the more serious adverse events [23–25]. Stimulation induced ADs are common during cortical mapping for epilepsy surgery, yet, despite decades of clinical observation, the cellular and network mechanisms underlying their generation remain areas of active investigation [26]. However, it is generally agreed that a disruption in excitatory-inhibitory balance results in the hyperexcitable state associated with these phenomena. During stimulation, the inhibitory drive on the post-synaptic neurons is increased, resulting in hyperpolarization. However, upon termination of the stimulus train, a phenomenon known as "post-inhibitory rebound excitation" can occur [27]. This rebound depolarization leads to a strong excitatory discharge in the primary neurons and may be one of the cellular mechanisms responsible for the generation of ADs. Using the same implantable device described here, this pattern of inhibition during stimulation, followed by strong excitatory bursts upon stimulus termination, has been observed in LFPs, chronically recorded from the sheep hippocampus [19].

This study indicates that AD activity in the ACC could be a biomarker for the likelihood of seizures. The relationship between the generation of ADs and the initial appearance of clinical seizures in these patients is unclear. Their recurrent seizures developed during a period where DBS was delivered in a continuous (not cycled) manner and could more likely be related to a kindling-like phenomenon. In animal models, classical kindling typically involves the application of periodic subthreshold stimuli to evoke network synchronization, which gradually induces long-lasting neuronal changes that eventually lead to spontaneous seizures. However, other kindling models employ higher level, more continuous stimulation, above the AD threshold, and result in a more rapid induction of epileptogenesis [28].

Stimulation with the cycle mode, rather than the prolonged continuous stimulation, has been proposed to reduce the risk of seizures [29]. Moreover, in an attempt to minimize the likelihood of ADs, a stimulation cycle that was slowly ramped off, rather than stopped abruptly, was evaluated in these patients. When stimulation was delivered using this pattern, the typical post-stimulation burst of spiking activity was not observed, even at intensity levels above those that earlier produced ADs, using the sensing capability of the implanted brain-machine-interface. Due to the large stimulus artifact, it was not possible to conclusively determine whether any spiking/ictal activity was present during stimulation. However, the stimulation pattern with a ramp appeared to avoid the generation of ADs following stimulation, possibly due to a reduction in the post-inhibitory rebound. Importantly, it has allowed for two patients to achieve long-term seizure freedom and pain relief.

Future systems would benefit from continuous monitoring of neural activity. Our data suggests that the likelihood of AD occurrence can fluctuate depending on the functional state of the stimulated network at that time [16]. This may explain why the LFPs recorded from patient 3 did not show ADs when using a stimulus without a ramp, which also suggests the importance of long-term monitoring and adaptive algorithms. Implanted sensing-enabled interfaces have the capability to chronically monitor for AD/ictal types of activity, based on spectral characteristics, and options to reduce or turn off stimulation, if detected [21,30]. Moreover, the sensing-enabled interfaces could be easily automated and run in the background and allow for more automated processing in the future. This type of closed-loop approach may potentially minimize or prevent stimulation-induced adverse events, such as those observed in this study.

New applications of DBS of new targets for therapy delivery continue to be explored; however, in most cases, the default stimulation parameters selected are based upon those that have been effective in the currently approved movement disorder therapies [31]. When exploring new therapies, a commonly accepted concept in neuromodulation is that side-effects and adverse events can be eliminated by simply turning off the device. However, the patients then fail to obtain symptomatic relief, and, as illustrated in this case series, there is a potential for sustained adverse events, even in the absence of stimulation. This caveat is important, as several recent large trials of DBS for new indications have not yielded positive outcomes [32–34] and a post-hoc review questioned whether stimulation parameters were adequately dosed. The unique opportunity to directly observe stimulation effects on the implanted structure [35,36] or neural network [37–39] targeted for therapy provided by sensing-enabled systems may usher in a new era, where DBS programming is informed by objective electrophysiological measures in conjunction with clinical observations, hopefully leading to safer and more effective therapies.

5. Conclusions

The events of unforeseen consequences following ACC DBS serve as a clarion call to those working in the field of neuromodulation. This report revealed that use of sensing-enabled systems could help to understand relationship between ACC stimulation and side-effects (seizures in these series), suggesting sensing-enabled techniques have the potential to advance safer brain stimulation therapies, especially in novel targets.

Author Contributions: Conceptualization, T.J.D. and T.Z.A.; methodology, Y.H., B.C., T.J.D., and T.Z.A.; formal analysis, Y.H.; investigation, Y.H., B.C., T.J.D., and T.Z.A.; data curation, Y.H.; writing—original draft preparation, Y.H., B.C., A.L.G., T.J.D., and T.Z.A.; writing—review and editing, Y.H., B.C., A.L.G., T.J.D., and T.Z.A.; visualization, Y.H.; supervision, T.J.D. and T.Z.A.; project administration, T.J.D. and T.Z.A.

Funding: This research received no external funding.

Acknowledgments: The authors thank Gaetano Leogrande, Paul Stypulkowski, and Scott Stanslaski for their help in data collection and manuscript review for technical accuracy. A.L.G. is supported by the National Institute for Health Research (NIHR) Oxford Biomedical Research Centre, based at Oxford University Hospitals NHS Trust and the University of Oxford.

Conflicts of Interest: T.J.D. is a former employee of Medtronic, Public Limited Company (PLC), which manufactures the neurostimulation system used in this work, and he has intellectual property in the areas of neural sensing, adaptive stimulation algorithms, and risk mitigations. B.C. is a current employee of Abbott, PLC, which manufactures a neurostimulation system. However, his contribution to the work presented in this manuscript was completed while he was employed by the University of Oxford. The implanted sensing-enabled neurostimulation system was donated by Medtronic and used on humanitarian grounds.

References

1. Lozano, A.M.; Lipsman, N.; Bergman, H.; Brown, P.; Chabardes, S.; Chang, J.W.; Matthews, K.; McIntyre, C.C.; Schlaepfer, T.E.; Schulder, M.; et al. Deep brain stimulation: Current challenges and future directions. *Nat. Rev. Neurol.* **2019**, *15*, 148–160. [CrossRef] [PubMed]

2. Hariz, M.I.; Blomstedt, P.; Zrinzo, L. Deep brain stimulation between 1947 and 1987: The untold story. *Neurosurg. Focus* **2010**, *29*, E1. [CrossRef] [PubMed]

3. Foltz, E.L.; White, L.E., Jr. Pain "relief" by frontal cingulumotomy. *J. Neurosurg.* **1962**, *19*, 89–100. [CrossRef] [PubMed]

4. Hassenbusch, S.J.; Pillay, P.K.; Barnett, G.H. Radiofrequency cingulotomy for intractable cancer pain using stereotaxis guided by magnetic resonance imaging. *Neurosurgery* **1990**, *27*, 220. [CrossRef] [PubMed]

5. Viswanathan, A.; Harsh, V.; Pereira, E.A.C.; Aziz, T.Z. Cingulotomy for medically refractory cancer pain. *Neurosurg. Focus* **2013**, *35*, E1. [CrossRef]

6. Cohen, R.A.; Kaplan, R.F.; Moser, D.J.; Jenkins, M.A.; Wilkinson, H. Impairments of attention after cingulotomy. *Neurology* **1999**, *53*, 819. [CrossRef] [PubMed]

7. Ochsner, K.N.; Kosslyn, S.M.; Cosgrove, G.; Cassem, E.H.; Price, B.H.; Nierenberg, A.A.; Rauch, S.L. Deficits in visual cognition and attention following bilateral anterior cingulotomy. *Neuropsychoogy* **2001**, *39*, 219–230. [CrossRef]

8. Boccard, S.G.; Fitzgerald, J.J.; Pereira, E.A.; Moir, L.; Van Hartevelt, T.J.; Kringelbach, M.L.; Green, A.L.; Aziz, T.Z. Targeting the affective component of chronic pain: A case series of deep brain stimulation of the anterior cingulate cortex. *Neurosurgery* **2014**, *74*, 628–635. [CrossRef]

9. Pereira, E.A.; Green, A.L.; Aziz, T.Z. Deep brain stimulation for pain. *Handbook Clin. Neurol.* **2013**, *116*, 277–294.

10. Maslen, H.; Cheeran, B.; Pugh, J.; Pycroft, L.; Boccard, S.; Prangnell, S.; Green, A.L.; FitzGerald, J.; Savulescu, J.; Aziz, T. Unexpected Complications of Novel Deep Brain Stimulation Treatments: Ethical Issues and Clinical Recommendations. *Neuromodulation* **2018**, *21*, 135–143. [CrossRef]

11. Blume, W.T.; Jones, D.C.; Pathak, P. Properties of after-discharges from cortical electrical stimulation in focal epilepsies. *Clin. Neurophysiol.* **2004**, *115*, 982–989. [CrossRef] [PubMed]

12. Gollwitzer, S.; Hopfengärtner, R.; Rössler, K.; Müller, T.; Olmes, D.G.; Lang, J.; Köhn, J.; Onugoren, M.D.; Heyne, J.; Schwab, S.; et al. Afterdischarges elicited by cortical electric stimulation in humans: When do they occur and what do they mean? *Epilepsy Behav.* **2018**, *87*, 173–179. [CrossRef] [PubMed]

13. Wieser, H.G.; Bancaud, J.; Talairach, J.; Bonis, A.; Szikla, G. Comparative Value of Spontaneous and Chemically and Electrically Induced Seizures in Establishing the Lateralization of Temporal Lobe Seizures. *Epilepsia* **1979**, *20*, 47–59. [CrossRef] [PubMed]

14. Lesser, R.P.; Kim, S.H.; Beyderman, L.; Miglioretti, D.L.; Webber, W.R.S.; Bare, M.; Cysyk, B.; Krauss, G.; Gordon, B. Brief bursts of pulse stimulation terminate afterdischarges caused by cortical stimulation. *Neurology* **1999**, *53*, 2073. [CrossRef]

15. Lesser, R.P.; Lüders, H.; Klem, G.; Dinner, D.S.; Morris, H.H.; Hahn, J. Cortical Afterdischarge and Functional Response Thresholds: Results of Extraoperative Testing. *Epilepsia* **1984**, *25*, 615–621. [CrossRef] [PubMed]

16. Lesser, R.P.; Lee, H.W.; Webber, W.R.S.; Prince, B.; Crone, N.E.; Miglioretti, D.L. Short-term variations in response distribution to cortical stimulation. *Brain* **2008**, *131*, 1528–1539. [CrossRef] [PubMed]

17. Lee, H.W.; Webber, W.; Crone, N.; Miglioretti, D.L.; Lesser, R.P. When is electrical cortical stimulation more likely to produce afterdischarges? *Clin. Neurophysiol.* **2010**, *121*, 14–20. [CrossRef]

18. Stypulkowski, P.; Stanslaski, S.; Denison, T.; Giftakis, J. Chronic Evaluation of a Clinical System for Deep Brain Stimulation and Recording of Neural Network Activity. *Ster. Funct. Neurosurg.* **2013**, *91*, 220–232. [CrossRef]

19. Stypulkowski, P.H.; Stanslaski, S.R.; Jensen, R.M.; Denison, T.J.; Giftakis, J.E. Brain Stimulation for Epilepsy—Local and Remote Modulation of Network Excitability. *Brain Stimul.* **2014**, *7*, 350–358. [CrossRef]

20. Rouse, A.G.; Stanslaski, S.R.; Cong, P.; Jensen, R.M.; Afshar, P.; Ullestad, D.; Gupta, R.; Molnar, G.F.; Moran, D.W.; Denison, T.J.; et al. A Chronic Generalized Bi-directional Brain-Machine Interface. *J. Neural Eng.* **2011**, *8*, 036018. [CrossRef]

21. Stanslaski, S.; Afshar, P.; Cong, P.; Giftakis, J.; Stypulkowski, P.; Carlson, D.; Linde, D.; Ullestad, D.; Avestruz, A.-T.; Denison, T. Design and Validation of a Fully Implantable, Chronic, Closed-Loop Neuromodulation Device With Concurrent Sensing and Stimulation. *IEEE Trans. Neural Syst. Rehabil. Eng.* **2012**, *20*, 410–421. [CrossRef] [PubMed]

22. Denison, T.J.; Afshar, P.; Stanslaski, S.R. Simultaneous Physiological Sensing and Stimulation with Saturation Detection. Google Patents 2016. Available online: https://patents.patsnap.com/v/US10080898-simultaneous-physiological-sensing-and-stimulation-with-saturation-detection.html (accessed on 25 September 2018).

23. Fontaine, D.; Hamani, C.; Lozano, A. Efficacy and safety of motor cortex stimulation for chronic neuropathic pain: Critical review of the literature. *J. Neurosurg.* **2009**, *110*, 251–256. [CrossRef] [PubMed]

24. Klein, M.M.; Treister, R.; Raij, T.; Pascual-Leone, A.; Park, L.; Nurmikko, T.; Lenz, F.; Lefaucheur, J.-P.; Lang, M.; Hallett, M.; et al. Transcranial magnetic stimulation of the brain: Guidelines for pain treatment research. *Pain* **2015**, *156*, 1601–1614. [CrossRef] [PubMed]

25. Cruccu, G.; Aziz, T.Z.; Garcia-Larrea, L.; Hansson, P.; Jensen, T.S.; Lefaucheur, J.-P.; Simpson, B.A.; Taylor, R.S. EFNS guidelines on neurostimulation therapy for neuropathic pain. *Eur. J. Neurol.* **2007**, *14*, 952–970. [CrossRef] [PubMed]

26. Kalamangalam, G.P.; Tandon, N.; Slater, J.D. Dynamic mechanisms underlying afterdischarge: A human subdural recording study. *Clin. Neurophysiol.* **2014**, *125*, 1324–1338. [CrossRef] [PubMed]

27. Llinás, R. The intrinsic electrophysiological properties of mammalian neurons: Insights into central nervous system function. *Science* **1988**, *242*, 1654–1664. [CrossRef]

28. Morales, J.C.; Roncagliolo, M.; Fuenzalida, M.; Wellmann, M.; Nualart, F.J.; Bonansco, C.; Álvarez-Ferradas, C. A new rapid kindling variant for induction of cortical epileptogenesis in freely moving rats. *Front. Cell. Neurosci.* **2014**, *8*, 200. [CrossRef]

29. De Ridder, D.; De Mulder, G.; Verstraeten, E.; Van Der Kelen, K.; Sunaert, S.; Smits, M.; Kovacs, S.; Verlooy, J.; Van De Heyning, P.; Moller, A.R. Primary and Secondary Auditory Cortex Stimulation for Intractable Tinnitus. *ORL* **2006**, *68*, 48–55. [CrossRef] [PubMed]

30. Afshar, P.; Khambhati, A.; Stanslaski, S.; Carlson, D.; Jensen, R.; Linde, D.; Dani, S.; Lazarewicz, M.; Cong, P.; Giftakis, J.; et al. A translational platform for prototyping closed-loop neuromodulation systems. *Front. Neural. Circuits* **2012**, *6*, 117. [CrossRef] [PubMed]

31. Miocinovic, S.; Somayajula, S.; Chitnis, S.; Vitek, J.L. History, Applications, and Mechanisms of Deep Brain Stimulation. *JAMA Neurol.* **2013**, *70*, 163. [CrossRef]

32. Dougherty, D.D.; Rezai, A.R.; Carpenter, L.L.; Howland, R.H.; Bhati, M.T.; O'Reardon, J.P.; Eskandar, E.N.; Baltuch, G.H.; Machado, A.D.; Kondziolka, D.; et al. A Randomized Sham-Controlled Trial of Deep Brain Stimulation of the Ventral Capsule/Ventral Striatum for Chronic Treatment-Resistant Depression. *Boil. Psychiatry* **2015**, *78*, 240–248. [CrossRef] [PubMed]

33. Lozano, A.M.; Fosdick, L.; Chakravarty, M.M.; Leoutsakos, J.-M.; Munro, C.; Oh, E.; Drake, K.E.; Lyman, C.H.; Rosenberg, P.B.; Anderson, W.S.; et al. A Phase II Study of Fornix Deep Brain Stimulation in Mild Alzheimer's Disease. *J. Alzheimer's Dis.* **2016**, *54*, 777–787. [CrossRef] [PubMed]

34. Holtzheimer, P.E.; Husain, M.M.; Lisanby, S.H.; Taylor, S.F.; A Whitworth, L.; McClintock, S.; Slavin, K.V.; Berman, J.; McKhann, G.M.; Patil, P.G.; et al. Subcallosal cingulate deep brain stimulation for treatment-resistant depression: A multisite, randomised, sham-controlled trial. *Lancet Psychiatry* **2017**, *4*, 839–849. [CrossRef]

35. Quinn, E.J.; Blumenfeld, Z.; Velisar, A.; Koop, M.M.; Shreve, L.A.; Trager, M.H.; Hill, B.C.; Kilbane, C.; Henderson, J.M.; Bronte-Stewart, H. Beta oscillations in freely moving Parkinson's subjects are attenuated during deep brain stimulation. *Mov. Disord.* **2015**, *30*, 1750–1758. [CrossRef] [PubMed]

36. Velisar, A.; Syrkin-Nikolau, J.; Blumenfeld, Z.; Trager, M.H.; Afzal, M.F.; Prabhakar, V.; Bronte-Stewart, H. Dual threshold neural closed loop deep brain stimulation in Parkinson disease patients. *Brain Stimul.* **2019**, *12*, 868–876. [CrossRef] [PubMed]

37. Van Gompel, J.J.; Klassen, B.T.; Worrell, G.A.; Lee, K.H.; Shin, C.; Zhao, C.Z.; Brown, D.A.; Goerss, S.J.; Kall, B.A.; Stead, M. Anterior nuclear deep brain stimulation guided by concordant hippocampal recording. *Neurosurg. Focus* **2015**, *38*, E9. [CrossRef] [PubMed]

38. Shute, J.B.; Okun, M.S.; Opri, E.; Molina, R.; Rossi, P.J.; Martinez-Ramirez, D.; Foote, K.D.; Gunduz, A. Thalamocortical network activity enables chronic tic detection in humans with Tourette syndrome. *NeuroImage Clin.* **2016**, *12*, 165–172. [CrossRef]

39. Swann, N.C.; De Hemptinne, C.; Miocinovic, S.; Qasim, S.; Ostrem, J.L.; Galifianakis, N.B.; Luciano, M.S.; Wang, S.S.; Ziman, N.; Taylor, R.; et al. Chronic multisite brain recordings from a totally implantable bidirectional neural interface: Experience in 5 patients with Parkinson's disease. *J. Neurosurg.* **2018**, *128*, 605–616. [CrossRef]

Article

Neurostimulation for Intractable Chronic Pain

Timothy R. Deer [1,*], Sameer Jain [2], Corey Hunter [3] and Krishnan Chakravarthy [4]

[1] Spine and Nerve Center of the Virginias, Charleston, VA 25301, USA
[2] Pain Treatment Centers of America, Little Rock, AR 72205, USA; paindocsj@gmail.com
[3] Ainsworth Institute of Pain Management, New York, NY 10022, USA; chunter@ainpain.com
[4] Department of Anesthesiology and Pain Medicine, University of California San Diego Health Sciences, San Diego, CA 92037, USA; kvchakravarthy@ucsd.edu
* Correspondence: doctdeer@aol.com; Tel.: +1-3043476141

Received: 11 December 2018; Accepted: 21 January 2019; Published: 24 January 2019

Abstract: The field of neuromodulation has seen unprecedented growth over the course of the last decade with novel waveforms, hardware advancements, and novel chronic pain indications. We present here an updated review on spinal cord stimulation, dorsal root ganglion stimulation, and peripheral nerve stimulation. We focus on mechanisms of action, clinical indications, and future areas of research. We also present current drawbacks with current stimulation technology and suggest areas of future advancements. Given the current shortage of viable treatment options using a pharmacological based approach and conservative interventional therapies, neuromodulation presents an interesting area of growth and development for the interventional pain field and provides current and future practitioners a fresh outlook with regards to its place in the chronic pain treatment paradigm.

Keywords: neuromodulation; neurostimulation; spinal cord stimulation; dorsal root ganglion stimulation; peripheral nerve stimulation; chronic pain

---

1. Introduction

Pain is an unpleasant sensory and emotional experience that involves complex processes of neuronal signaling in the peripheral nervous system (PNS) and the central nervous system (CNS). Chronic pain may be defined as pain persistent for more than 3–6 months [1]. For decades, chronic pain conditions continue to pose an immense burden on the economy and society in the form of healthcare expenditures and years lived with disability (YLD). Lower back pain alone has been the leading cause of YLDs for the past three decades [2]. In the United States, the healthcare expenditure secondary to chronic pain conditions in the year 2010 was estimated to be $560–$635 billion dollars [3]. This cost was more than the combined expenditure on heart diseases and diabetes mellitus. Globally, 10% of adults are diagnosed with chronic pain conditions each year [4]. Considering the vast amount of suffering caused by chronic pain conditions, international resolutions were made to make access to adequate pain therapy a human right [5,6]. Unfortunately, chronic pain has been known to be notoriously resistant to conventional medical management (CMM) [7,8]. This drove physicians to resort to using opioid therapy to manage chronic pain, inadvertently leading to what we now known as the "Opioid crisis". There have been an increasing number of deaths involving the overuse of prescription opioids [9]. Unfortunately, this upward trend has continued and remains a major cause of morbidity and mortality. Physicians are now on the constant lookout for opioid sparing therapies to manage chronic pain, and neuromodulation may be the answer. Neuromodulation is the process of inhibition, stimulation, modification, or therapeutic alteration of activity in the CNS, PNS, or the autonomic nervous system (ANS), with the use of electricity. In this review, we will highlight some of the important targets for neuromodulation therapy, their mechanism of action, and the evidence to support their use in the treatment of chronic intractable pain conditions.

2. Materials and Methods

The authors performed a search of the current literature using PubMed, Google scholar, Cochrane, Embase, and Medline database. In addition, the review of current scientific meetings, proceedings, and regulatory approvals were used to focus on modern advancements in the field. We selected and cited the major peer-reviewed publications supporting the use of neuromodulation for the management of various painful conditions.

2.1. Spinal Cord Stimulation (SCS)

The most simplistic description of Spinal Cord Stimulation (SCS) may be the application of electricity to the dorsal columns of the spinal cord to modulate/manipulate the pain signals carried by the ascending pain pathways to the brain, and hence is also known as dorsal column stimulation (DCS). The concept of SCS derives its inspiration from the landmark "Gate control theory of Pain" proposed by Melzack and Wall in 1965 [10]. This theory postulated the existence of a "Gate" in the dorsal horn of the spinal cord controlling the traffic of neuronal impulses from the sensory afferent neurons to the higher centers in the brain responsible for pain perception. Aβ fibers (responsible for carrying the non-nociceptive stimuli) and C fibers (responsible for carrying the painful stimuli) form synapses with the projection neurons of the spinothalamic tract on the dorsal horn of the spinal cord, which are responsible for the transmission of pain signals to the brain. According to the "gate control theory", stimulation of the Aβ fibers in the same region as the C fibers can result in the closure of the "gate", and thus resulting in blocking the transmission of pain impulses. In the spinal cord, these fibers are conveniently segregated from the motor fibers and are in an accessible location, making the dorsal columns a desirable target for stimulation. Based on this theory, Shealy et al. implanted the first dorsal column stimulator in 1967 for the treatment of pain [11]. However, several decades of research has shown that the mechanism of SCS in the treatment of pain is much more complex and continues to elude us.

2.2. Parameters of Stimulation

In order to understand the new stimulation paradigms and their mechanisms of action, it is critical to get a better understanding of how the delivery of charge to the spinal cord is manipulated. The three main parameters of stimulation include amplitude, pulse width, and frequency. The basic unit of electrical stimulation in neuromodulation is the "pulse", which consists of a sustained delivery of a specific amount of current amplitude (measured in milliamperes, mA) for a specific amount of time (pulse width, measured in microseconds, μs). The amount of charge delivered with each pulse is equivalent to the product of amplitude and pulse width, whereas, frequency determines the number of pulses delivered per second. Thus, alteration in the values of these parameters determines the amount of current (amount of charge delivered per second) that is delivered to the neurons. Therefore, narrow pulse widths require high amplitudes to activate the neuron or axon whereas wider pulse widths need lower amplitudes. The amount of charge needed to activate an axon in vivo depends upon the size, myelination, and the distance from the stimulation source. Primarily amplitude with some contribution from pulse width determines the number of fibers recruited and results in a perceived increase or decrease in the intensity and/or area of paresthesia sensation. Frequency of stimulation influences how often a neuron fires in response to a stimulus.

2.3. SCS Waveforms and Their Mechanisms of Action

2.3.1. Conventional/Tonic SCS

Conventional/tonic stimulation was the only stimulation paradigm available until few years ago and continues to be the most frequently used in clinical practice. This stimulation paradigm is characterized by low frequency (40–100 Hz), high amplitude (3.6–8.5 mA), and pulse widths ranging between 300–600 μs. It results in the delivery of "high charge" per pulse resulting in the perceived "paresthesias" by the patient. This SCS program has demonstrated superiority over conventional medical management (CMM) strategies in the treatment of several neuropathic (e.g., complex regional pain syndrome (CRPS), diabetic neuropathy, neuropathic limb pain, etc.) and mixed neuropathic (e.g., failed back surgery syndrome (FBSS)) chronic pain conditions (Table 1).

Even though "gate control theory" formed the basis of spinal cord stimulation, but it was primitive and failed to explain why SCS was ineffective in the treatment of acute nociceptive pain. Several theories have been proposed since then explaining the mechanism of action of SCS. Hyperexcitability of the wide-dynamic range (WDR) neurons in the dorsal horn (DH) of the spinal cord has been demonstrated in neuropathic pain states [12]. In animal models, SCS frequencies around 50 Hz have shown to induce release of inhibitory neurotransmitters like GABA resulting in inhibition of the WDR hyperexcitability [13,14]. It has also been suggested that SCS results in release of acetylcholine and its action on muscarinic M4 receptors may be responsible for its analgesic effects [15]. Furthermore, evidence indicates that the pain reduction with SCS may be secondary to stimulation-induced release of serotonin, adenosine, and noradrenaline [16]. Recent evidence suggests the involvement of supraspinal circuitry in mediating the analgesic effects of SCS [17,18]. However, the exact mechanism for the analgesic effects of SCS is still not clear.

Table 1. Indications and outcomes of conventional SCS.

Study	Indication	Study Design	Methods	Outcome Measures	Results	Conclusion
Kemler et al. [19]	CRPS	Randomized trial	CRPS patients assigned in a 2:1 ratio to SCS + PT group ($n = 36$) & PT group ($n = 18$). 24 of 36 patients underwent permanent implant of SCS device.	VAS, GPE, functional status, health-related quality of life	Intention-to-treat analysis showed significant reductions in pain at 6 m in SCS + PT group ($p < 0.001$). Improvements in GPE also observed in SCS group.	SCS can reduce pain in carefully selected CRPS patients.
Harke et al. [20]	Sympathetically maintained CRPS	Prospective trial	CRPS patients underwent SCS implant, and pain intensity was estimated during SCS free intervals of 45 min every 3 m vs. under treatment.	VAS, pain disability index, reduction in pain medication	Improvements in VAS during treatment vs. SCS free intervals ($p < 0.01$). Reduction in pain meds during treatment ($p < 0.01$).	Functional status, quality of life, and pain medication usage can be improved with use of SCS in sympathetically mediated CRPS.
North et al. [21]	FBSS	Randomized controlled trial	50 FBSS patients randomized to SCS and reoperation. If results of randomized treatment unsatisfactory, patient could crossover to alternative.	Self-reported pain relief, patient satisfaction, crossover to alternative procedure	Among 45 patients available for follow up, SCS (9 of 19) was more successful than reoperation (3 of 26 patients) ($p < 0.01$). (5 of 24 in SCS group) vs. (14 of 26 in reoperation group) crossed over ($p = 0.02$).	SCS is more effective than reoperation in patients with persistent radicular pain after spine surgery.
Kumar et al. [22]	FBSS/Neuropathic limb pain	Multicenter randomized controlled trial	100 FBSS patients with predominant leg pain of neuropathic radicular origin randomized to SCS + CMM group vs. CMM alone group and followed for 6 m.	1^0 outcome—$\geq$50% pain relief in the legs. 2^0 outcome—improvement in back and leg pain, health-related quality of life, functional capacity, use of pain medications	In the intention-to-treat analysis at 6m, 48% SCS patients ($n = 24$) & 9% CMM patients ($n = 4$) achieved 1^0 outcome. SCS + CMM group also achieved the 2^0 outcomes significantly more than the CMM alone group ($p < 0.05$ for all comparisons).	SCS is superior to CMM in the treatment of limb pain of neuropathic origin in patients with prior lumbosacral surgery.
De vos et al. [23]	PDN	Multicenter randomized controlled trial	60 PDN patients refractory to conventional medical therapy were randomized in 2:1 ratio to best conventional medical practice (with SCS) or without (control) SCS group and followed at regular intervals.	EuroQoL 5D, SF-MPQ, VAS	At 6m follow up, average VAS decreased from 73 (baseline) to 31 in SCS group ($p < 0.001$); VAS remained unchanged at 67 in control group ($p = 0.97$). SF-MPQ and EuroQoL 5D also improved significantly in the SCS group.	SCS therapy significantly reduced pain and improved quality of life in patients with PDN.
Van beek et al. [24]	PDN	Prospective two-center clinical trial	48 patients with PDN were treated with SCS and followed for 5 years.	NRS score for pain, PGIC, and treatment success (50% reduction of NRS score or significant PGIC)	Patients showed significant improvements in all outcome measures at the follow-up visits. Treatment success was observed in 55% of patients after 5 years, and 80% of patients with permanent implant continued to use their SCS device.	SCS is successful in alleviating pain in patients with PDN.

SCS—Spinal Cord Stimulation, VAS—Visual analog scale, GPE—Global perceived effect, PDN—Painful diabetic neuropathy SF-MPQ—Short-form Mcgill pain questionnaire, NRS—Numeric rating scale, PGIC—Patient's global impression of change, PT—Physical Therapy, EuroQol 5D- EuroQol five dimensions questionnaire.

2.3.2. High Frequency (HF) SCS

Recent clinical investigations have emphasized the importance of the way energy is delivered to the neural structures in neuromodulation therapies [25]. This has led the scientists in the last few years to focus on the development of new waveforms and stimulation paradigms. High frequency (10 khz) stimulation with a pulse width at 30 µs and amplitude ranging between 1–5 mA is among the most recent developments made on that front [26]. This stimulation therapy has shown superiority over conventional/tonic stimulation in the treatment of chronic low back pain and improving quality of life in a randomized controlled trial (RCT) (Table 2). However, there is still no evidence to support the use of HF SCS over conventional SCS stimulation in the treatment of chronic neuropathic limb pain.

The mechanisms by which HF stimulation results in analgesia are not fully understood, but several working hypotheses have been proposed. One of the theories is that it induces a depolarization blockade (a local reversible block), where propagating action potentials are blocked by HF stimulation [27]. Another hypothesis is that HF stimulation can induce a desynchronization of neural signals from clusters of neurons firing in synchrony. This results in pseudospontaneous or stochastic neural activity, where firing becomes individualized such that each unit is firing at its own rate and pattern [28,29]. "Membrane integration" has also been suggested as a possible mechanism of action for HF SCS, where multiple impulses reaching a neuron within a certain time frame may depolarize it and fire an action potential although every individual impulse is insufficient [30].

Table 2. Comparative studies between conventional/tonic & High Frequency (10 khz) SCS.

Study	Indication	Study Design	Methods	Outcome Measures	Results	Conclusion
Kapural et al. [26]	Chronic Intractable back and leg pain	Randomized controlled trial	198 subjects with back and leg pain, randomized in 1:1 ratio to HF SCS (>10 khz) or conventional SCS. Of these 171 passed trial and received permanent implant.	1^0 outcome—$\geq$50% pain relief in the back	At 3 months, In HF group 84.5% were responders for back pain (vs. 43.8% for tonic SCS) and 83.1% were responders for leg pain (vs. 55.5% for tonic SCS); ($p < 0.001$). Superiority of HF stimulation was sustained through a 12-month period.	HF stimulation was better than tonic stimulation for treatment of chronic intractable back and leg pain.
Kapural et al. [31]	Chronic Intractable back and leg pain	Randomized controlled trial	198 subjects with back and leg pain, randomized in 1:1 ratio to HF SCS (>10 khz) or conventional SCS. Of these 171 passed the trial and received permanent implant.	1^0 outcome—$\geq$50% pain relief in the back	At 24-months follow up, more subjects continued to be responders to HF stimulation than conventional SCS (back pain-76.5% vs. 49.3%, leg pain-72.9% vs. 49.3%; $p < 0.001$). Also back and leg pain decreased to a greater degree with HF stimulation than tonic SCS ($p < 0.001$).	HF (10 khz) stimulation was better than tonic stimulation for treatment of chronic intractable back and leg pain.
Amirdelfan et al. [32]	Chronic Intractable back and leg pain	Randomized controlled trial	198 subjects with back and leg pain, randomized in 1:1 ratio to HF SCS (>10 khz) or conventional SCS. Of these 171 passed the trial and received permanent implant. QOL and functional measures were collected up to 12 months.	ODI, GAF, CGIC, PSQI, SF-MPQ-2	At 12 months follow up; ODI-69.6% subjects were classified into lower disability category with HF (vs. 55.1% with tonic SCS; $p = 0.01$). Subjects had a more significant improvement in GAF scores in HF group vs. tonic SCS (14 vs. 6.5, respectively; $p < 0.01$). Significant improvements were seen in continuous, intermittent, and neuropathic pain in HF group vs. tonic SCS on the SF-MPQ-2 scale. However, no difference was observed on the affective disorders subscale. Significant improvements were also seen in the HF group on CGIC and PSQ1 scales compared to tonic SCS.	High frequency (10 khz) stimulation was better than tonic stimulation in improving quality of life and functional outcomes in patients with chronic intractable back and leg pain.

SCS—Spinal Cord Stimulation, High frequency—HF, ODI—Oswestry disability index, GAF—Global assessment of functioning, CGIC—Clinical global impression of change, PSQI—Pittsburgh sleep quality index, SF-MPQ—Short-form Mcgill pain questionnaire, QOL—Quality of life.

2.3.3. Burst SCS

This novel SCS waveform (series of five 1000 μs pulses delivered at 500 Hz followed by a repolarization pulse, with each series repeated at 40 Hz) is another SCS paradigm that has proven superior to conventional SCS in the treatment of lumbosacral component of pain (Table 3). This waveform is reported to mimic the firing patterns of endogenous neurons responsible for encoding aspects of pain signaling in the thalamus [33–35]. De ridder et al., on the basis of "source localized EEG" findings also postulated that burst SCS, via modulation of the medial spinothalamic pathway, could activate cortical areas involved in the modulation of pain perception [36]. Thus, making it capable of engaging both spinal and supraspinal pathways in both an anterograde and retrograde fashion as well as those medial and lateral supraspinal pathways. Another hypothesis is that burst firing may be capable of disrupting the synchronous burst firing of the high threshold fibers and inhibiting the activation directly related to pain perception [37]. Even though burst SCS has been shown to be more effective than the tonic SCS stimulation in the treatment of mixed neuropathic pain syndromes like FBSS [38,39], there is not enough evidence to support its superiority in the treatment of pure neuropathic pain states like diabetic neuropathy, CRPS, etc. A prospective observational study was conducted to compare burst SCS vs. HF SCS on a small cohort of 14 FBSS patients who underwent trials with burst ($n = 8$) and HF SCS ($n = 6$) [40,41]. Even though no significant difference was found in the effectiveness to treat the low back pain, burst SCS was slightly superior (not statistically significant) to HF SCS in treating the leg pain component.

Table 3. Comparative studies between conventional/tonic & burst SCS.

Study	Indication	Study Design	Methods	Outcome Measures	Results	Conclusion
De ridder et al. [38]	Intractable neuropathy/FBSS/diabetic neuropathy	Retrospective analysis	Retrospective analysis of 102 patients who previously received SCS was performed. These were divided into two groups—the first group included patients who became failures to tonic stimulation and others who continue to respond. Both groups switched to burst SCS and followed up.	NRS pain scores, amount of responders.	It was reported that almost 25% of the patients were non-responders to conventional SCS and out of that, 63% responded to burst. Also, 95% who responded to tonic stimulation reported further improvement with burst SCS.	Burst was better than tonic stimulation and can also rescue non- responders.
Deer et al. [39]	FBSS/Persistent radicular pain	Randomized controlled trial	100 patients with a successful trial with tonic SCS randomized to receive tonic vs. burst stimulation for the first 12 weeks and the other stimulation mode for the next 12 weeks. Subjects then used the stimulation mode of their choice and were followed for a year.	1^0 endpoint- assessment of VAS score (tonic vs. burst)	Intention-to-treat analysis was used to estimate the difference in overall VAS scores, which showed burst was superior to tonic stimulation ($p < 0.017$). Significantly more subjects (70.8%) preferred burst over tonic stimulation ($p < 0.001$).	Burst stimulation was superior to tonic stimulation for the treatment of chronic pain.
Demartini et al. [42]	FBSS/persistent radicular pain	Multicenter observational study	23 patients underwent 2 weeks of tonic stimulation followed by 2 weeks of burst stimulation.	1^0 outcome-reduction of pain in the back and the legs. 2^0 EuroQol-5D, PCS	Tonic stimulation reduced leg pain ($p < 0.05$), the burst mode added an extra pain reduction (ΔNRS 1.2 ± 1.5) ($p < 0.01$). Both stimulation paradigms failed to reduce back pain ($p = 0.29$) Secondary outcomes were achieved with both stimulation paradigms.	Burst stimulation was more successful than tonic stimulation in the treatment of leg pain.

SCS—Spinal Cord Stimulation, NRS—Numeric rating scale, VAS—Visual analog scale, PCS—Pain catastrophizing scale.

2.4. Closed-Loop Spinal Cord Stimulation

Even though SCS therapy has numerous proven benefits, it does seem to have certain flaws. One of the major issues with conventional SCS (open-loop) systems is the need for manual adjustment of stimulation current to maintain coverage during postural changes. The position of SCS electrodes in relation to the dorsal column of the spinal cord is dynamic and varies with postural changes [43,44]. This results in unwanted side effects and sometimes loss of therapy. For example, a decrease in the distance between the electrodes and the spinal cord may result in activation of unwanted fibers, which may result in unwanted or uncomfortable paresthesias, muscle twitching, and cramping. Conversely, an increase in the distance between the spinal cord and the SCS electrodes may result in loss of therapy. Loss of therapy/efficacy with spinal cord stimulation has been a concern among pain physicians for a long period of time. Previous studies have demonstrated that effective pain control with SCS decreases over time [45–47]. A prospective study demonstrated a linear increase in VAS scores after one and two years of follow up ($p = 0.03$). However, VAS scores were still significantly lower than pre-SCS therapy. A systematic literature review demonstrated successful pain relief in 62% patients at one year with SCS therapy, whereas the success rate dropped to 53% and 35% patients at five and 10 year follow up respectively [48]. Some of the possible causes that have been speculated for this loss of therapy are progression in the underlying disease, change in paresthesia coverage, device migration/malfunction, changes in microenvironment of the electrode leading to high impedances [49].

Closed-loop SCS was developed to neutralize the side effects encountered with postural changes. This stimulation therapy measures individual evoked compound action potential (ECAP) and uses them as a feedback control mechanism to automatically maintain desired dorsal column fiber recruitment levels. The ECAP amplitude at which patient experiences optimal pain relief is set as the reference and the feedback algorithm alters the input current to maintain it constant. Russo et al. published the preliminary results of a prospective, multicenter, single-arm study showing effectiveness and safety of the closed-loop SCS system in the treatment of leg and low back pain [50]. In this study, 51 patients with chronic low back and leg pain underwent a trial with closed-loop SCS system. Thirty-six patients later underwent permanent implantation and were followed for six months. Significant reductions ($\geq$80%) in pain were observed in 70.4% (back pain) and 56.5% (leg pain) patients at the 3-month interval, and 64.3% (back pain) and 60.9% (leg pain) patients at 6-month follow up. Statistically significant improvements in mean BPI (Brief pain inventory), EQ-5D-5L, ODI (Oswestry disability index), and PSQI (Pittsburgh sleep quality index) were also observed at both time points.

2.5. Dorsal Root Ganglion (DRG) Stimulation

While the utility and efficacy of traditional SCS is well-established in the literature, the therapy is not without its shortcomings. These deficiencies range from paresthesias in unwanted areas and waning relief over time to position-related changes in stimulation intensity and the inability to capture focal areas like the foot and pelvic region [51–57]. Perhaps one of the most pressing concerns surrounding the traditional SCS was its inability to provide sustained pain relief in patients with chronic, focal neuropathic pain despite being considered by many to be the "treatment of choice" for such conditions. Long-term data from a prospective study suggested that treating CRPS with SCS and physical therapy may be no better than physical therapy alone after 2-years [58]. These shortcomings led scientists and clinicians to look for new targets within the central nervous system as means to improve upon the therapy that is neuromodulation; one such target was the DRG.

The DRG was long thought of as a passive neural structure that acted solely as a support structure, facilitating communication between peripheral and central nervous systems [59]. The idea that it played any relevant role in the development or maintenance of chronic neuropathic pain had not been elucidated until recently. Current evidence suggests that the DRG, itself, is directly responsible for the development of neuropathic pain through "hyperexcitability" and "spontaneous ectopic firing" of those neurons contained within the DRG [59,60], two processes mainly responsible for central sensitization and allodynia (the hallmarks of CRPS). When one also takes into account the DRG's role in the modulation of sensory processing and nociceptive pain as well as predictable anatomical location and scarcity of CSF that would otherwise deflect energy [61–63], the DRG appeared to be an ideal target for stimulation that could potentially bridge the gap for neurostimulation as a whole.

2.6. Physiology

When a peripheral nerve become injured on inflamed, there are a number of changes that occur within the actual DRG:

Gene expression [64]
Microglial cells [65]
Ion channels & current [59]
Chemokines [59]
Ectopic Firing [59,60]
Hyperexcitability [59,60]

Even more interesting is the role the DRG plays in the filtering of transmissions from the peripheral nervous system into the central. The cell bodies of the neurons located with the DRG possess a t-junction that give them the ability to filter action potentials and pool stimuli from the periphery until a certain threshold is achieved before opening up and allowing the signal into the central nervous system [66–68].

Taking into account the sheer variety of relevant processes now known occur at the level of the DRG, targeting it for neuromodulation appeared to be a logical conclusion. DRG stimulation is believed to impact pain by applying a variety of effects the processes thought to not only develop chronic pain, but maintain it [69–71]:

- Activation of supraspinal centers and the deactivation of hyperexcitability of wide-dynamic range (WDR) neurons located within the dorsal horn;
- Upstream/downstream effects causing stabilization of peripheral nociceptor sensitization, vasodilation, release of neuromodulators in the dorsal horn, and activation of WDR neurons;
- Theorized normalization of gene expression within the DRG and spinal cord;
- Augmentation of T-junction "low pass filter" thus reducing propagation of action potential to the dorsal horn;
- Decreased hyperexcitability of neurons within the DRG by down regulation of abnormal; Na^+ channels, up-regulation of K^+ channels and restoration of normal calcium flow;
- Stabilizing microglia releasing cytokines (TNF-α, chemokines, nerve growth factors, interleukins, interferons, etc.).

2.7. Evidence for Efficacy

The primary indication for DRGs is focal neuropathic pain, namely CRPS. The early pilot studies by Deer, Grigsby, and Liem not only proved the concept that stimulating the DRG was viable, but also that superior levels of pain relief not typically attainable with conventional SCS were reported (Table 4). In 2012, Deer et al. reported on a prospective study of 10 patients trialed with DRG for 3–7 days; complaints included discogenic pain, low back pain with radicular symptoms, DPPN, PHN (Post-herpetic neuralgia) and neuropathic chest wall pain [56]. This pilot study showed a 70% reduction in pain in the majority of patients with commensurate decrease in opioid consumption.

In 2013, Liem et al. reported on the results of a prospective, 1-year study of 32 patients treated with DRGS [72]. The patients in this study included CRPS, Failed Back Surgery Syndrome (FBSS), chronic post-surgical pain, PHN, spinal stenosis, discogenic pain and radicular pain. Overall pain reduction was 56% with 52% of the subjects reporting >50% improvement in pain. More importantly, was the 80% reduction in foot pain, an area of the body that traditionally been difficult to treat with SCS. Additionally, the authors reported that the patients denied posture-dependent fluctuations in paresthesias commonly associated with SCS.

In 2017, Deer et al. reported the results of the ACCURATE study, a multicenter, randomized, controlled trial of 152 subjects with CRPS treated with DRGS and followed out to 1-year (control group received traditional dorsal column SCS) [73]. The study showed that DRGS was statistically superior to traditional SCS with 74.2% of the DRGS group reporting 50% or more pain relief at 1-year, compared to the control group's 53%. In addition, patients treated with DRGS reported 81.4%–86% decrease in VAS compared to 48.1%–70.2% decreases in the control group.

Table 4. Evidence for efficacy for DRG stimulation.

Study	Indication	Study Design	Methods	Outcome Measures	Results	Conclusion
Deer et al. [56]	Chronic intractable neuropathic pain of trunk and/or limbs	Prospective, multicenter, single arm, pilot study	10 subjects underwent trial with Dorsal root ganglion stimulation device and were followed up for 3–7 days.	Daily VAS scores, perceived % of pain relief at the final visit	Average pain reduction between baseline and final follow up visit was 70 + 32% ($p = 0.0007$). All subjects achieved pain relief in the desired specific regions of the body.	Authors concluded DRG could be a viable target for neurostimulation for the treatment of chronic intractable pain.
Liem et al. [72]	Chronic intractable neuropathic pain of trunk/limb and/or sacral region	Prospective, Multicenter study	32 subjects with successful trial with DRG stimulation underwent permanent implantation of the device. Patients were followed up for 6 months.	VAS, % of pain relief at follow up, improvements in quality of life (EQ-5D), mood, function.	At 6 month follow up, overall pain reduction was 56%; 52% patients had >50% pain relief. Improvements were seen in all other outcome measures.	Neuromodulation of DRG was effective in the treatment of chronic intractable neuropathic pain conditions. It is able to provide paresthesia coverage in areas such as foot, which were difficult to treat with traditional SCS.
Deer et al. [73]	CRPS	Multicenter, randomized controlled trial	152 subjects randomized in a 1:1 ratio to receive DRG stimulation vs. traditional SCS and were followed up at 3, 6, 9, and 12 months	1^0 end point—>50% reduction in VAS scores at 3 month follow up 2^0 end point- positional effects on paresthesia intensity	The percentage of subjects with >50% pain relief was greater in DRG arm (81.2%) vs. SCS arm (55.7%, $p < 0.001$) at 3 months. Subjects in DRG arm reported less postural variation in paresthesia ($p < 0.001$).	DRG stimulation was more effective and provided less postural variation as compared to conventional SCS.

DRG—Dorsal root ganglion, SCS—Spinal cord stimulation, VAS—Visual analog scale, EQ-5D—EuroQol five dimensions questionnaire.

Since the inception DRGS, a number of manuscripts have been published on a variety of unique and novel uses that have proven truly groundbreaking, not only for neuromodulation, but the field pain medicine as a whole. Syndromes that had proven to be recalcitrant to most well-accepted pain treatments, including SCS, now had published evidence showing they could potentially be treated with DRG stimulation:

- Post-herniorrhaphy neuralgia [74,75]
- Post-amputee pain and phantom limb pain [76,77]
- Post surgical chest wall pain (i.e., post-mastectomy & post-thoractomy pain) [78–80]
- Chronic pelvic pain [57]
- Knee pain after total joint arthroplasty [74,81]
- Post-herpetic neuralgia [56,72,74]
- Diabetic peripheral neuropathy [23,82–84]

In 2018, Deer et al. published a "Best Practices" manuscript on the use of DRGS, along with a grading of the available evidence as well as recommendations on its use for various indications [85]. Aside from CRPS (which has Level-I evidence to support) most the other indications had varying degrees of Level-II evidence with recommendation grades ranging between A to B (extremely recommendable to recommendable).

2.8. Peripheral Nerve Stimulation

An area of growing interest in the field of neuromodulation has been peripheral nerve stimulation (PNS). With the ability to limit the amount of energy dispersion by using focalized current this area of therapy provides an unprecedented opportunity to treat a multitude of chronic pain disorders. In 1999, the first peripheral nerve leads were placed percutaneously to manage intractable headaches [86]. This has been expanded to include modulation of visceral, neuropathic, cardiac, abdominal, back, and facial pain. Though there are many studies to deduce PNS mechanism of action that validate the Wall and Melzack gate control theory, it has been postulated that PNS is used as a method of orthodromic stimulation of non-nociceptive Aβ nerve fibers. Activation of these fibers results in excitation of respective dorsal horn inter-neurons that are involved in processing and transmitting nociceptive information via peripheral Aβ and C nerve fibers. Thus, non-painful stimulation of the peripheral nerve territory results in decreased pain signals [87]. Studies have suggested an acute modulation of the local microenvironment with down-regulation of neurotransmitters and endorphins in addition to local inflammatory mediators may also be a critical piece on how PNS may be effective in treating chronic pain. Other potential methods of pain modulation could result from reducing ectopic discharges in addition to reducing Wallerian degeneration.

2.9. Summary of Clinical Indications

There is growing evidence of the use of peripheral nerve stimulation in a variety of clinical indications that include plexus injuries, focal mononeuropathy, post-amputation pain, back pain, sacroiliac joint pain, headache, facial pain, arm and limb pain. Prior studies have shown that there are good outcomes from PNS on median, ulnar, sciatic, ilioinguinal, and genito-femoral nerves [88–92]. Specific data also supports use of PNS following stimulation of brachial plexus and lumbar plexus with reduction in neuropathic pain, allodynia and restoration of normal tactile sensation following respective plexus injuries [93,94].

With regards to post-amputation pain, Rauck et al. have shown that following two weeks of home trial nine responders reported reductions across several variables, including mean daily worst post-amputation pain, average residual limb pain, average phantom limb pain, residual limb pain interference, phantom limb pain interference, and Pain Disability Index up to four weeks following the end of stimulation. These positive findings were counterbalanced by minor decreases in the Beck Depression Inventory scores with little to no change in pain medication use.

Other approaches have also looked at peripheral nerve field stimulation (PNFS) where the electrode contact point is placed at the area of pain but not in direct contact with the nerve. The direct neural target using this form of peripheral field stimulation also targets Aβ nerve fibers consistent with peripheral nerve stimulation. Klomstein et al. evaluated the long-term efficacy and safety of PNFS in lower back pain in 105 patients at 1, 3, and 6 months post-implantation. They observed a stable decrease in pain at 6 months. Mean VAS score at baseline was VAS 7.9 (SD 1.38) and 4.7 (SD 1.99) at six-month follow up ($p < 0.01$). Statistically significant improvements were also seen across other parameters, including the Oswestry Disability Questionnaire, the Becks Depression Inventory, and the Short Form-12 item Health survey. Of the enrolled subjects 9.6% of the subjects experienced complications requiring surgical intervention [95]. Guentchev et al. also recently reported on the utility of PNS in managing sacroiliac joint (SIJ) pain [96]. This 12-patient study using eight pole electrode placed parallel to the SIJ joint showed at two weeks' post-implant, subjects reported an average Oswetry Disability Index ODI reduction from 57% to 32% and VAS from 9 to 2.1. International Patient Satisfaction Index (IPSI) was 1.1. At six months, the mean ODI was 34% ($p = 0.0006$), VAS was 3.8 ($p < 0.0001$) and IPSI was 1.9. At 12 months, mean averages for 6 of 7 patients were ODI 21% ($p < 0.0005$), VAS 1.7 ($p < 0.0001$), and IPSI 1.3 [96].

With regards to headache and facial pain there has been numerous studies looking at the benefits for PNS on migraines. The ONSTIM study was a prospective single-blind 66 patient randomized study that showed a 39% response in the simulation group and 6% response in the pre-set simulation group based on an responder rate of greater than 50% or VAS improvement of 3 [97]. Dodick et al. presented 12-month data evaluating the use of PNS of the occipital nerves for patients with chronic migraine [98]. Headache days were significantly decreased by 6.7 (±8.4) days in the intent-to-treat (ITT) population ($p < 0.001$) and by 7.7 (±8.7) days in the intractable chronic migraine (ICM) population ($p < 0.001$). Excellent or good headache relief was also reported by almost two thirds of the ITT population and close to 70% of the ICM population. The study reported 183 device/procedure-related adverse events, of which 18 (8.6%) required hospitalization and 85 (40.7%) required surgical intervention [98]. Cluster headaches have been shown to respond to Sphenopalatine ganglion (SPG) stimulation as well stimulation of the pterygopalatine fossa [97].

There has also been good benefits reported from PNS for refractory sub-acromial impingement syndrome (SIS) [99], post-stroke shoulder pain [100], and post-traumatic brachial plexus trauma refractory to medical and surgical management [101].

Though an independent topic of its own right there has been considerable recent developments of implantable and portable vagus nerve stimulators that have shown to modulate nociception in addition to efficacy in treatment of refractory epilepsy and depression [102–107]. Clinical areas that have shown good effect have been trigeminal allodynia, fibromyalgia, chronic pelvic pain and headaches.

Several more recent advances in peripheral nerve stimulation technology has resulted in more improved compliance and ease of use. Deer et al. conducted an eight-patient trial targeting the median nerve for alleviating neuropathic pain using a novel stimrouter system with wireless battery to lead connectivity. They observed both pain reduction throughout the 5-day treatment period and reduced oral opioid consumption with no significant or unexpected adverse events [108]. This was followed by Deer et al. publishing a randomized double-blinded multicenter trial of 147 patients that showed that patients receiving active stimulation achieved a statistically significantly higher response rate of 38% vs. the 10% rate found in the Control group ($p = 0.0048$). Specifically, the treatment group achieved a mean pain reduction of 27.2% from Baseline to Month 3 of follow-up compared to a 2.3% reduction in the Control group ($p < 0.0001$). The study did not report any adverse events [109].

Potential reported adverse events mainly included lead migration, hardware issues (i.e., battery failure, lead of extension disconnection, programmer malfunction, IPG migration and malfunction). Other reported events include subcutaneous hematomas, seromas, skin erosions, pain and numbness at the IPG site, allergic reactions to surgical material, headache and muscle cramping [98,101,108–110].

3. Discussion

We have attempted to present a comprehensive review of the current areas of neuromodulation advances and their potential uses in various chronic pain pathologies. Spinal cord stimulation (SCS) is a well established modality to effectively control the pain of neuropathic origin. Its efficacy and safety has been demonstrated in several randomized controlled trials. For several decades conventional SCS was the only stimulation paradigm available to patients. Even though treatment with this modality showed great results, it was not without its shortcomings including but not limited to failure of therapy, unwanted paresthesias, and development of tolerance. In the last few years, research was focused on manipulation of SCS parameters to meet the physiological needs of the patients. Development of Burst SCS program was a step in this direction where stimulation mimics the natural neuronal firing patterns. It was found to be more effective than conventional SCS in the treatment of low back pain. Similarly, development of high frequency stimulation therapy, which is presumed to act via induction of depolarization blockade/desynchronization of neuronal signals has also shown superiority over conventional SCS in the management of chronic low back pain. These two new stimulation paradigms also provide patients with an option of paresthesia free stimulation, which may be preferred by some patients. However, the more recent research in the field of spinal cord stimulation is focused on altering the therapy to individual needs. Development of closed loop SCS is a step in this direction to mitigate the effects of positional changes and development of tolerance to the therapy.

While dorsal column stimulation has shown great promise, alternative technology outside the dorsal column focused on concentrating current in the dorsal root ganglion or targeting individual nerves as demonstrated through peripheral nerve stimulation has shown growing promise. Questions related to reduced energy dispersion, focused targeted therapy, and potential effects of these various dorsal column and peripheral nerve stimulator on the neuroimmune axis presents exciting future research. In addition, given the growing understanding of various waveforms and their respective effects on the medial and lateral pain pathways may provide more insight into mechanism of action and help to tailor more appropriate therapy for each individual patient.

While constant efforts are being made to advance the field of neuromodulation, a challenge that is consistently faced by researchers is the inability to produce ideal study designs. Secondary to the intrinsic nature of therapy, it is nearly impossible to blind the patient, physician, and the programmer to produce reliable test results. Also, use of "sham-effect" raises the ethical concerns of subjecting the patient to the risks of an interventional pain procedure with no benefit.

4. Conclusions

These are times of advancement in the field of bioelectrical medicine. With this progress comes new responsibilities for those involved in this revolution. The responsibilities include a commitment to improving efficacy, mitigating complications, and finding new innovations that may continue to evolve the progress that has been made to date. This should be done with a commitment to ethics and patient safety, and with a curiosity that inspires new ideas and discoveries.

Author Contributions: All authors contributed equally to this manuscript. Each author served as a primary author.

Funding: This research received no external funding.

Acknowledgments: None.

Conflicts of Interest: Timothy R. Deer is a consultant for Abbott, Vertos Medical, Flowonix, Axonics, SpineThera, Saluda Medical, Nalu Medical, Vertiflex and Ethos. He has Stock Options with Bioness, Vertos, Axonics, SpineThera, Saluda, Nalu, Vertiflex, and Cornerloc. Paid research with Abbott: Saluda, Mainstay, and Vertiflex. Patent Pending with Abbott.

References

1. Treede, R.D.; Rief, W.; Barke, A.; Aziz, Q.; Bennett, M.I.; Benoliel, R.; Cohen, M.; Evers, S.; Finnerup, N.B.; First, M.B.; et al. A classification of chronic pain for ICD-11. *Pain* **2015**, *156*, 1003–1007. [CrossRef] [PubMed]
2. Disease, G.B.D.; Injury, I.; Prevalence, C. Global, regional, and national incidence, prevalence, and years lived with disability for 328 diseases and injuries for 195 countries, 1990–2016: A systematic analysis for the Global Burden of Disease Study 2016. *Lancet* **2017**, *390*, 1211–1259.
3. Gaskin, D.J.; Richard, P. The economic costs of pain in the United States. *J. Pain* **2012**, *13*, 715–724. [CrossRef] [PubMed]
4. Goldberg, D.S.; McGee, S.J. Pain as a global public health priority. *BMC Public Health* **2011**, *11*, 770. [CrossRef] [PubMed]
5. World Medical Association. WMA Resolution on the Access to Adequate Pain Treatment. (Adopted by the 62nd WMA General Assembly, Montevideo, Uruguay, October 2011). Available online: http://www.wma.net/en/30publications/10policies/p2/index.html (accessed on 6 November 2018).
6. International Association for the Study of Pain and Delegates of International Pain Summit 2010. Declaration that Access to Pain Management Is a Fundamental Human Right. 2010. Available online: http://www.iasp-pain.org/Content/NavigationMenu/Advocacy/DeclarationofMontr233al/default.htm (accessed on 6 November 2018).
7. Dworkin, R.H.; O'connor, A.B.; Backonja, M.; Farrar, J.T.; Finnerup, N.B.; Jensen, T.S.; Kalso, E.A.; Loeser, J.D.; Miaskowski, C.; Nurmikko, T.J.; et al. Pharmacologic management of neuropathic pain: Evidence-based recommendations. *Pain* **2007**, *132*, 237–251. [CrossRef] [PubMed]

8. Moulin, D.E.; Clark, A.J.; Gilron, I.; Ware, M.A.; Watson, C.P.; Sessle, B.J.; Coderre, T.; Morley-Forster, P.K.; Stinson, J.; Boulanger, A.; et al. Pharmacological management of chronic neuropathic pain—Consensus statement and guidelines from the Canadian Pain Society. *Pain Res. Manag.* **2007**, *12*, 13–21. [CrossRef] [PubMed]

9. Annual Surveillance Report of Drug-Related Risks and Outcomes—United States. Available online: https://www.cdc.gov/drugoverdose/pdf/pubs/2017-cdc-drug-surveillance-report.pdf (accessed on 19 August 2017).

10. Melzack, R.; Wall, P.D. Pain mechanisms: A new theory. *Science* **1965**, *150*, 971–979. [CrossRef] [PubMed]

11. Shealy, C.N.; Mortimer, J.T.; Reswick, J.B. Electrical inhibition of pain by stimulation of the dorsal columns: Preliminary clinical report. *Anesth Analg.* **1967**, *46*, 489–491. [CrossRef]

12. Yakhnitsa, V.; Linderoth, B.; Meyerson, B.A. Spinal cord stimulation attenuates dorsal horn neuronal hyperexcitability in a rat model of mononeuropathy. *Pain* **1999**, *79*, 223–233. [CrossRef]

13. Cui, J.G.; O'Connor, W.T.; Ungerstedt, U.; Linderoth, B.; Meyerson, B.A. Spinal cord stimulation attenuates augmented dorsal horn release of excitatory amino acids in mononeuropathy via a GABAergic mechanism. *Pain* **1997**, *73*, 87–95. [CrossRef]

14. Stiller, C.O.; Cui, J.G.; O'Connor, W.T.; Brodin, E.; Meyerson, B.A.; Linderoth, B. Release of gamma-aminobutyric acid in the dorsal horn and suppression of tactile allodynia by spinal cord stimulation in mononeuropathic rats. *Neurosurgery* **1996**, *39*, 367–374. [CrossRef] [PubMed]

15. Schechtmann, G.; Song, Z.; Ultenius, C.; Meyerson, B.A.; Linderoth, B. Cholinergic mechanisms involved in the pain relieving effect of spinal cord stimulation in a model of neuropathy. *Pain* **2008**, *139*, 136–145. [CrossRef] [PubMed]

16. Linderoth, B.; Gazelius, B.; Franck, J.; Brodin, E. Dorsal column stimulation induces release of serotonin and substance P in the cat dorsal horn. *Neurosurgery* **1992**, *31*, 289–296. [CrossRef] [PubMed]

17. Saade, N.E.; Barchini, J.; Tchachaghian, S.; Chamaa, F.; Jabbur, S.J.; Song, Z.; Meyerson, B.A.; Linderoth, B. The role of the dorsolateral funiculi in the pain relieving effect of spinal cord stimulation: A study in a rat model of neuropathic pain. *Exp. Brain Res.* **2015**, *233*, 1041–1052. [CrossRef] [PubMed]

18. Barchini, J.; Tchachaghian, S.; Shamaa, F.A.; Jabbur, S.J.; Meyerson, B.A.; Song, Z.; Linderoth, B.; Saade, N.E. Spinal segmental and supraspinal mechanisms underlying the pain-relieving effects of spinal cord stimulation: An experimental study in a rat model of neuropathy. *Neuroscience* **2012**, *215*, 196–208. [CrossRef] [PubMed]

19. Kemler, M.A.; Barendse, G.A.; Van Kleef, M.; de Vet, H.C.; Rijks, C.P.; Furnee, C.A.; Van Den Wildenberg, F.A. Spinal cord stimulation in patients with chronic reflex sympathetic dystrophy. *N. Engl. J. Med.* **2000**, *343*, 618–624. [CrossRef] [PubMed]

20. Harke, H.; Gretenkort, P.; Ladleif, H.U.; Rahman, S. Spinal cord stimulation in sympathetically maintained complex regional pain syndrome type I with severe disability. A prospective clinical study. *Eur. J. Pain* **2005**, *9*, 363–373. [CrossRef] [PubMed]

21. North, R.B.; Kidd, D.H.; Farrokhi, F.; Piantadosi, S.A. Spinal cord stimulation versus repeated lumbosacral spine surgery for chronic pain: A randomized, controlled trial. *Neurosurgery* **2005**, *56*, 98–106. [CrossRef]

22. Kumar, K.; Taylor, R.S.; Jacques, L.; Eldabe, S.; Meglio, M.; Molet, J.; Thomson, S.; O'callaghan, J.; Eisenberg, E.; Milbouw, G.; et al. Spinal cord stimulation versus conventional medical management for neuropathic pain: A multicentre randomised controlled trial in patients with failed back surgery syndrome. *Pain* **2007**, *132*, 179–188. [CrossRef]

23. de Vos, C.C.; Meier, K.; Zaalberg, P.B.; Nijhuis, H.J.; Duyvendak, W.; Vesper, J.; Enggaard, T.P.; Lenders, M.W. Spinal cord stimulation in patients with painful diabetic neuropathy: A multicentre randomized clinical trial. *Pain* **2014**, *155*, 2426–2431. [CrossRef]

24. van Beek, M.; Geurts, J.W.; Slangen, R.; Schaper, N.C.; Faber, C.G.; Joosten, E.A.; Dirksen, C.D.; van Dongen, R.T.; van Kuijk, S.M.; van Kleef, M. Severity of Neuropathy Is Associated With Long-term Spinal Cord Stimulation Outcome in Painful Diabetic Peripheral Neuropathy: Five-Year Follow-up of a Prospective Two-Center Clinical Trial. *Diabetes Care* **2018**, *41*, 32–38. [CrossRef] [PubMed]

25. Pope, J.E.; Falowski, S.; Deer, T.R. Advanced waveforms and frequency with spinal cord stimulation: Burst and high-frequency energy delivery. *Expert Rev. Med. Devices* **2015**, *12*, 431–437. [CrossRef] [PubMed]

26. Kapural, L.; Yu, C.; Doust, M.W.; Gliner, B.E.; Vallejo, R.; Sitzman, B.T.; Amirdelfan, K.; Morgan, D.M.; Brown, L.L.; Yearwood, T.L.; et al. Novel 10-kHz High-frequency Therapy (HF10 Therapy) Is Superior to Traditional Low-frequency Spinal Cord Stimulation for the Treatment of Chronic Back and Leg Pain: The SENZA-RCT Randomized Controlled Trial. *Anesthesiology* **2015**, *123*, 851–860. [CrossRef] [PubMed]

27. Kilgore, K.L.; Bhadra, N. Reversible nerve conduction block using kilohertz frequency alternating current. *Neuromodulation* **2014**, *17*, 242–254. [CrossRef] [PubMed]

28. Litvak, L.M.; Smith, Z.M.; Delgutte, B.; Eddington, D.K. Desynchronization of electrically evoked auditory-nerve activity by high-frequency pulse trains of long duration. *J. Acoust. Soc. Am.* **2003**, *114*, 2066–2078. [CrossRef] [PubMed]

29. Rubinstein, J.T.; Wilson, B.S.; Finley, C.C.; Abbas, P.J. Pseudospontaneous activity: Stochastic independence of auditory nerve fibers with electrical stimulation. *Hear Res.* **1999**, *127*, 108–118. [CrossRef]

30. Reilly, J.P.; Freeman, V.T.; Larkin, W.D. Sensory effects of transient electrical stimulation–evaluation with a neuroelectric model. *IEEE Trans. Biomed. Eng.* **1985**, *32*, 1001–1011. [CrossRef] [PubMed]

31. Kapural, L.; Yu, C.; Doust, M.W.; Gliner, B.E.; Vallejo, R.; Sitzman, B.T.; Amirdelfan, K.; Morgan, D.M.; Yearwood, T.L.; Bundschu, R.; et al. Comparison of 10-kHz High-Frequency and Traditional Low-Frequency Spinal Cord Stimulation for the Treatment of Chronic Back and Leg Pain: 24-Month Results From a Multicenter, Randomized, Controlled Pivotal Trial. *Neurosurgery* **2016**, *79*, 667–677. [CrossRef]

32. Amirdelfan, K.; Yu, C.; Doust, M.W.; Gliner, B.E.; Morgan, D.M.; Kapural, L.; Vallejo, R.; Sitzman, B.T.; Yearwood, T.L.; Bundschu, R.; et al. Long-term quality of life improvement for chronic intractable back and leg pain patients using spinal cord stimulation: 12-month results from the SENZA-RCT. *Qual. Life Res.* **2018**, *27*, 2035–2044. [CrossRef]

33. Radhakrishnan, V.; Tsoukatos, J.; Davis, K.D.; Tasker, R.R.; Lozano, A.M.; Dostrovsky, J.O. A comparison of the burst activity of lateral thalamic neurons in chronic pain and non-pain patients. *Pain* **1999**, *80*, 567–575. [CrossRef]

34. Lenz, F.A.; Garonzik, I.M.; Zirh, T.A.; Dougherty, P.M. Neuronal activity in the region of the thalamic principal sensory nucleus (ventralis caudalis) in patients with pain following amputations. *Neuroscience* **1998**, *86*, 1065–1081. [CrossRef]

35. Emmers, R. Thalamic mechanisms that process a temporal pulse code for pain. *Brain Res.* **1976**, *103*, 425–441. [CrossRef]

36. De Ridder, D.; Plazier, M.; Kamerling, N.; Menovsky, T.; Vanneste, S. Burst spinal cord stimulation for limb and back pain. *World Neurosurg.* **2013**, *80*, 642–649. [CrossRef] [PubMed]

37. Ahmed, S.; Yearwood, T.; de Ridder, D.; Vanneste, S. Burst and High Frequency Stimulation: Underlying Mechanism of Action. *Expert Rev. Med. Devices* **2018**, *15*, 61–70. [CrossRef] [PubMed]

38. De Ridder, D.; Lenders, M.W.; De Vos, C.C.; Dijkstra-Scholten, C.; Wolters, R.; Vancamp, T.; Van Looy, P.; Van Havenbergh, T.; Vanneste, S. A 2-center comparative study on tonic versus burst spinal cord stimulation: Amount of responders and amount of pain suppression. *Clin. J. Pain* **2015**, *31*, 433–437. [CrossRef] [PubMed]

39. Deer, T.; Slavin, K.V.; Amirdelfan, K.; North, R.B.; Burton, A.W.; Yearwood, T.L.; Tavel, E.; Staats, P.; Falowski, S.; Pope, J.; et al. Success Using Neuromodulation With BURST (SUNBURST) Study: Results From a Prospective, Randomized Controlled Trial Using a Novel Burst Waveform. *Neuromodulation* **2018**, *21*, 56–66. [CrossRef] [PubMed]

40. Kinfe, T.M.; Pintea, B.; Link, C.; Roeske, S.; Güresir, E.; Güresir, Á.; Vatter, H. High Frequency (10 kHz) or Burst Spinal Cord Stimulation in Failed Back Surgery Syndrome Patients With Predominant Back Pain: Preliminary Data From a Prospective Observational Study. *Neuromodulation* **2016**, *19*, 268–275. [CrossRef]

41. Muhammad, S.; Roeske, S.; Chaudhry, S.R.; Kinfe, T.M. Burst or High-Frequency (10 kHz) Spinal Cord Stimulation in Failed Back Surgery Syndrome Patients With Predominant Back Pain: One Year Comparative Data. *Neuromodulation* **2017**, *20*, 661–667. [CrossRef]

42. Demartini, L.; Terranova, G.; Innamorato, M.A.; Dario, A.; Sofia, M.; Angelini, C.; Duse, G.; Costantini, A.; Leoni, M.L. Comparison of Tonic vs. Burst Spinal Cord Stimulation During Trial Period. *Neuromodulation* **2018**. [CrossRef]

43. Schade, C.M.; Schultz, D.M.; Tamayo, N.; Iyer, S.; Panken, E. Automatic Adaptation of Neurostimulation Therapy in Response to Changes in Patient Position: Results of the Posture Responsive Spinal Cord Stimulation (Prs) Research Study. *Pain Phys.* **2011**, *14*, 407–417.

44. Sun, T.F.; Morrell, M.J. Closed-Loop Neurostimulation: The Clinical Experience. *Neurotherapeutics* **2014**, *11*, 553–563. [CrossRef] [PubMed]

45. Alo, M.K.; Redko, V.; Charnov, J. Four Year Follow-up of Dual Electrode Spinal Cord Stimulation for Chronic Pain. *Neuromodulation* **2002**, *5*, 79–88. [CrossRef] [PubMed]

46. Forouzanfar, T.; Kemler, M.A.; Weber, W.E.; Kessels, A.G.; van Kleef, M. Spinal Cord Stimulation in Complex Regional Pain Syndrome: Cervical and Lumbar Devices Are Comparably Effective. *Br. J. Anaesth.* **2004**, *92*, 348–353. [CrossRef]

47. Mann, A.S.; Sparkes, E.; Duarte, R.V.; Raphael, J.H. Attrition with Spinal Cord Stimulation. *Br. J. Anaesth.* **2015**, *29*, 823–828. [CrossRef] [PubMed]

48. Turner, A.J.; Loeser, J.D.; Bell, K.G. Spinal Cord Stimulation for Chronic Low Back Pain: A Systematic Literature Synthesis. *Neurosurgery* **1995**, *37*, 1088–1095. [CrossRef] [PubMed]

49. Aiudi, M.C.; Dunn, R.Y.; Burns, S.M.; Roth, S.A.; Opalacz, A.; Zhang, Y.; Chen, L.; Mao, J.; Ahmed, S.U. Loss of Efficacy to Spinal Cord Stimulator Therapy: Clinical Evidence and Possible Causes. *Pain Phys.* **2017**, *20*, E1073–E1080.

50. Russo, M.; Cousins, M.J.; Brooker, C.; Taylor, N.; Boesel, T.; Sullivan, R.; Poree, L.; Shariati, N.H.; Hanson, E.; Parker, J. Effective Relief of Pain and Associated Symptoms With Closed-Loop Spinal Cord Stimulation System: Preliminary Results of the Avalon Study. *Neuromodulation* **2018**, *21*, 38–47. [CrossRef] [PubMed]

51. Frey, M.E.; Manchikanti, L.; Benyamin, R.M.; Schultz, D.M.; Smith, H.S.; Cohen, S.P. Spinal cord stimulation for patients with failed back surgery syndrome: A systematic review. *Pain Phys.* **2009**, *12*, 379–397.

52. North, R.B. Neural interface devices: Spinal cord stimulation technology. *Proc. IEEE* **2008**, *96*, 1108–1119. [CrossRef]

53. Turner, J.A.; Loeser, J.D.; Deyo, R.A.; Sanders, S.B. Spinal cord stimulation for patients with failed back surgery syndrome or complex regional pain syndrome: A systematic review of effectiveness and complications. *Pain* **2004**, *108*, 137–147. [CrossRef] [PubMed]

54. Cameron, T. Safety and efficacy of spinal cord stimulation for the treatment of chronic pain: A 20-year literature review. *J. Neurosurg.* **2004**, *100*, 254–267. [CrossRef] [PubMed]

55. Hunter, C.; Dave, N.; Diwan, S.; Deer, T. Neuromodulation of pelvic visceral pain: Review of the literature and case series of potential novel targets for treatment. *Pain Pract.* **2013**, *13*, 3–17. [CrossRef] [PubMed]

56. Deer, T.R.; Grigsby, E.; Weiner, R.L.; Wilcosky, B.; Kramer, J.M. A prospective study of dorsal root ganglion stimulation for the relief of chronic pain. *Neuromodulation* **2013**, *16*, 67–71. [CrossRef] [PubMed]

57. Hunter, C.W.; Yang, A. Dorsal Root Ganglion Stimulation for Chronic Pelvic Pain: A Case Series and Technical Report on a Novel Lead Configuration. *Neuromodulation* **2018**. [CrossRef] [PubMed]

58. Kemler, M.A.; de Vet, H.C.; Barendse, G.A.; van den Wildenberg, F.A.; van Kleef, M. Effect of spinal cord stimulation for chronic complex regional pain syndrome Type I: Five-year final follow-up of patients in a randomized controlled trial. *J. Neurosurg.* **2008**, *108*, 292–298. [CrossRef] [PubMed]

59. Krames, E.S. The dorsal root ganglion in chronic pain and as a target for neuromodulation: A review. *Neuromodulation* **2015**, *18*, 24–32. [CrossRef] [PubMed]

60. Sapunar, D.; Ljubkovic, M.; Lirk, P.; McCallum, J.B.; Hogan, Q.H. Distinct membrane effects of spinal nerve ligation on injured and adjacent dorsal root ganglion neurons in rats. *Anesthesiology* **2005**, *103*, 360–376. [CrossRef]

61. Hasegawa, T.; Mikawa, Y.; Watanabe, R.; An, H.S. Morphometric analysis of the lumbosacral nerve roots and dorsal root ganglia by magnetic resonance imaging. *Spine* **1996**, *21*, 1005–1009. [CrossRef]

62. Hasegawa, T.; An, H.S.; Haughton, V.M. Imaging anatomy of the lateral lumbar spinal canal. *Semin. Ultrasound* **1993**, *14*, 404–413. [CrossRef]

63. Devor, M. Unexplained Peculiarities of the Dorsal Root Ganglion. *Pain* **1999**, *6*, S27–S35. [CrossRef]

64. Herdegen, T.; Fiallos-Estrada, C.E.; Schmid, W.; Bravo, R.; Zimmermann, M. The transcription factors c-JUN, JUN D and CREB, but not FOS and KROX-24, are differentially regulated in axotomized neurons following transection of rat sciatic nerve. *Brain Res. Mol. Brain Res.* **1992**, *14*, 155–165. [CrossRef]

65. Hanani, M. Satellite glial cells in sensory ganglia: From form to function. *Brain Res. Mol. Brain Res.* **2005**, *48*, 457–476. [CrossRef] [PubMed]

66. Zhang, J.M.; Donnelly, D.F.; LaMotte, R.H. Patch clamp recording from the intact dorsal root ganglion. *J. Neurosci. Methods* **1998**, *79*, 97–103. [CrossRef]

67. Luscher, C.; Streit, J.; Quadroni, R.; Luscher, H.R. Action potential propagation through embryonic dorsal root ganglion cells in culture. I. Influence of the cell morphology on propagation properties. *J. Neurophysiol.* **1994**, *72*, 622–633. [CrossRef] [PubMed]

68. Wang, W.; Gu, J.; Li, Y.Q.; Tao, Y.X. Are voltage-gated sodium channels on the dorsal root ganglion involved in the development of neuropathic pain? *Mol. Pain* **2011**, *7*, 16. [CrossRef] [PubMed]

69. Croom, J.E.; Foreman, R.D.; Chandler, M.J.; Barron, K.W. Cutaneous vasodilation during dorsal column stimulation is mediated by dorsal roots and CGRP. *Am. J. Physiol.* **1997**, *272*, H950–H957. [CrossRef] [PubMed]

70. Lee, D.C. Modulation of bipolar neuron activity by extracellular electric field. *Neuroscience* **2011**. [CrossRef]

71. Koopmeiners, A.S.; Mueller, S.; Kramer, J.; Hogan, Q.H. Effect of electrical field stimulation on dorsal root ganglion neuronal function. *Neuromodulation* **2013**, *16*, 304–311. [CrossRef]

72. Liem, L.; Russo, M.; Huygen, F.J.; Van Buyten, J.P.; Smet, I.; Verrills, P.; Cousins, M.; Brooker, C.; Levy, R.; Deer, T.; et al. A Multicenter, Prospective Trial to Assess the Safety and Performance of the Spinal Modulation Dorsal Root Ganglion Neurostimulator System in the Treatment of Chronic Pain. *Neuromodulation* **2013**, *16*, 471–482. [CrossRef] [PubMed]

73. Deer, T.R.; Levy, R.M.; Kramer, J.; Poree, L.; Amirdelfan, K.; Grigsby, E.; Staats, P.; Burton, A.W.; Burgher, A.H.; Obray, J.; et al. Dorsal root ganglion stimulation yielded higher treatment success rate for complex regional pain syndrome and causalgia at 3 and 12 months: A randomized comparative trial. *Pain* **2017**, *158*, 669–681. [CrossRef] [PubMed]

74. Hunter, C.W.; Sayed, D.; Lubenow, T.; Davis, T.; Carlson, J.; Rowe, J.; Justiz, R.; McJunkin, T.; Deer, T.; Mehta, P.; et al. DRG FOCUS: A Multicenter Study Evaluating Dorsal Root Ganglion Stimulation and Predictors for Trial Success. *Neuromodulation* **2018**. [CrossRef] [PubMed]

75. Schu, S.; Gulve, A.; ElDabe, S.; Baranidharan, G.; Wolf, K.; Demmel, W.; Rasche, D.; Sharma, M.; Klase, D.; Jahnichen, G.; et al. Spinal cord stimulation of the dorsal root ganglion for groin pain-a retrospective review. *Pain Pract.* **2015**, *15*, 293–299. [CrossRef] [PubMed]

76. Eldabe, S.; Burger, K.; Moser, H.; Klase, D.; Schu, S.; Wahlstedt, A.; Vanderick, B.; Francois, E.; Kramer, J.; Subbaroyan, J. Dorsal Root Ganglion (DRG) Stimulation in the Treatment of Phantom Limb Pain (PLP). *Neuromodulation* **2015**, *18*, 610–616. [CrossRef]

77. Hunter, C.W.; Yang, A.; Davis, T. Selective Radiofrequency Stimulation of the Dorsal Root Ganglion (DRG) as a Method for Predicting Targets for Neuromodulation in Patients With Post Amputation Pain: A Case Series. *Neuromodulation* **2017**, *20*, 708–718. [CrossRef] [PubMed]

78. Breel, J.; Zuidema, X.; Schapendonk, J.; Lo, B.; Wille, F.F. Treatment of Chronic Post-surgical Pain with Dorsal Root Ganglion (drg) Stimulation (10701). *Neuromodulation* **2016**, *19*, e15.

79. Liem, L.; Nijhuis, H.; Huygen, F.J.; Gulve, A.; Sharma, M.L.; Patel, N.K.; Love-jones, S.; Green, A.L.; Baranidharan, G.; Bush, D.; Eldabe, S. treatment Of Chronic Post-surgical Pain (cpsp) Using Spinal Cord Stimulation (scs) Of The Dorsal Root Ganglion (drg) Neurostimulation: Results From Two Prospective Studies: Wip-0479. *Pain Pract.* **2014**, *14*, 3–4.

80. Espinet, A.J. Stimulation of Dorsal Root Ganglion (drg) for Chronic Post-surgical Pain (cpsp): A retrospective single centre case series from australia using targeted Spinal Cord Stimulation (scs). *Neuromodulation* **2015**, *18*, 84.

81. van Bussel, C.M.; Stronks, D.L.; Huygen, F.J. Successful treatment of intractable complex regional pain syndrome type I of the knee with dorsal root ganglion stimulation: A case report. *Neuromodulation* **2015**, *18*, 58–60. [CrossRef]

82. Schu, S.; Mutagi, H.; Kapur, S.; Guru, R.; Leljevahl, E.; Wahlstedt, A.; Eldabe, S. sustained pain relief in Painful Diabetic Neuropathy (pdn) achieved through targeted Spinal Cord Stimulation (scs): A retrospective case series. *Neuromodulation* **2015**, *18*, 82.

83. Eldabe, S.; Espinet, A.; Kang, P.; Liem, L.; Patel, N.; Tadros, M. Dorsal root ganglia stimulation for painful diabetic peripheral neuropathy: A preliminary report. *Neuromodulation* **2017**, *20*, e51.

84. Slangen, R.; Schaper, N.C.; Faber, C.G.; Joosten, E.A.; Dirksen, C.D.; van Dongen, R.T.; Kessels, A.G.; van Kleef, M. Spinal cord stimulation and pain relief in painful diabetic peripheral neuropathy: A prospective two-center randomized controlled trial. *Diabetes Care* **2014**, *37*, 3016–3024. [CrossRef] [PubMed]

85. Deer, T.R.; Pope, J.E.; Lamer, T.J.; Grider, J.S.; Provenzano, D.; Lubenow, T.R.; FitzGerald, J.J.; Hunter, C.; Falowski, S.; Sayed, D.; et al. The Neuromodulation Appropriateness Consensus Committee on Best Practices for Dorsal Root Ganglion Stimulation. *Neuromodulation* **2018**. [CrossRef] [PubMed]

86. Weiner, R.L.; Reed, K.L. Peripheral neurostimulation for control of intractable occipital neuralgia. *Neuromodulation* **1999**, *2*, 217–221. [CrossRef]

87. Papuc, E.; Rejdak, K. The role of neurostimulation in the treatment of neuropathic pain. *Ann. Agric. Environ. Med.* **2013**, *1*, 14–17.

88. De La Cruz, P.; Gee, L.; Walling, I.; Morris, B.; Chen, N.; Kumar, V.; Feustel, P.; Shin, D.S.; Pilitsis, J.G. Treatment of Allodynia by Occipital Nerve Stimulation in Chronic Migraine Rodent. *Neurosurgery* **2015**, *77*, 479–485. [CrossRef] [PubMed]

89. Stinson, L.W., Jr.; Roderer, G.T.; Cross, N.E.; Davis, B.E. Peripheral Subcutaneous Electrostimulation for Control of Intractable Post-operative Inguinal Pain: A Case Report Series. *Neuromodulation* **2001**, *4*, 99–104. [CrossRef] [PubMed]

90. Kothari, S.; Goroszeniuk, T. Percutaneous permanent electrode implantation to ulnar nerves for upper extremity chronic pain: 6 years follow up: 201. *Reg. Anesth. Pain Med.* **2006**, *31*, 16. [CrossRef]

91. Siddalingaiah, V.; Goroszeniuk, T.; Palmisani, S. Permanent percutaneous sciatic nerve stimulation for treatment of severe neuropathic Pain. Presented at the 13th World Congress on Pain, Montreal, QC, Canada, 29 August–2 September 2010.

92. Shetty, A.; Pang, D.; Mendis, V.; Sood, M.; Goroszeniuk, T. median Nerve Stimulation In Forearm For Treatment Of Neuropathic Pain Post Reimplantation Of Fingers: A Case Report. *Pain Pract.* **2012**, *12*, 343.

93. Goroszeniuk, T.; Kothari, S.C.; Hamann, W.C. Percutaneous implantation of a brachial plexus electrode for management of pain syndrome caused by a traction injury. *Neuromodulation* **2007**, *10*, 148–155. [CrossRef]

94. Petrovic, Z.; Goroszeniuk, T.; Kothari, S.; Padfield, N. Percutaneous lumbar plexus stimulation in the treatment of intractable pain. *Reg. Anesth. Pain Med.* **2007**, *32*, 11. [CrossRef]

95. Frahm, K.S.; Hennings, K.; Vera-Portocarrero, L.; Wacnik, P.W.; Morch, C.D. Muscle Activation During Peripheral Nerve Field Stimulation Occurs Due to Recruitment of Efferent Nerve Fibers, Not Direct Muscle Activation. *Neuromodulation* **2016**, *19*, 587–596. [CrossRef]

96. Kloimstein, H.; Likar, R.; Kern, M.; Neuhold, J.; Cada, M.; Loinig, N.; Ilias, W.; Freundl, B.; Binder, H.; Wolf, A.; et al. Peripheral nerve field stimulation (PNFS) in chronic low back pain: A prospective multicenter study. *Neuromodulation* **2014**, *17*, 180–187. [CrossRef] [PubMed]

97. Goroszeniuk, T.; Pang, D. Peripheral neuromodulation: A review. *Curr. Pain Headache Rep.* **2014**, *18*, 412. [CrossRef] [PubMed]

98. Dodick, D.W.; Silberstein, S.D.; Reed, K.L.; Deer, T.R.; Slavin, K.V.; Huh, B.; Sharan, A.D.; Narouze, S.; Mogilner, A.Y.; Trentman, T.L.; et al. Safety and efficacy of peripheral nerve stimulation of the occipital nerves for the management of chronic migraine: Long-term results from a randomized, multicenter, double-blinded, controlled study. *Cephalalgia* **2015**, *35*, 344–358. [CrossRef] [PubMed]

99. Wilson, R.D.; Harris, M.A.; Gunzler, D.D.; Bennett, M.E.; Chae, J. Percutaneous peripheral nerve stimulation for chronic pain in subacromial impingement syndrome: A case series. *Neuromodulation* **2014**, *17*, 771–776. [CrossRef] [PubMed]

100. Wilson, R.D.; Gunzler, D.D.; Bennett, M.E.; Chae, J. Peripheral nerve stimulation compared with usual care for pain relief of hemiplegic shoulder pain: A randomized controlled trial. *Am. J. Phys. Med. Rehabil.* **2014**, *93*, 17–28. [CrossRef] [PubMed]

101. Stevanato, G.; Devigili, G.; Eleopra, R.; Fontana, P.; Lettieri, C.; Baracco, C.; Guida, F.; Rinaldo, S.; Bevilacqua, M. Chronic post-traumatic neuropathic pain of brachial plexus and upper limb: A new technique of peripheral nerve stimulation. *Neurosurg. Rev.* **2014**, *37*, 473–479. [CrossRef]

102. Multon, S.; Schoenen, J. Pain control by vagus nerve stimulation: From animal to man and back. *Acta Neurol. Belg.* **2005**, *105*, 62–67.

103. Kirchner, A.; Birklein, F.; Stefan, H.; Handwerker, H.O. Left vagus nerve stimulation suppresses experimentally induced pain. *Neurology* **2000**, *55*, 1167–1171. [CrossRef]

104. Ness, T.J.; Randich, A.; Fillingim, R.; Faught, R.E.; Backensto, E.M. Left vagus nerve stimulation suppresses experimentally induced pain. *Neurology* **2001**, *56*, 985–986. [CrossRef]

105. Ness, T.J.; Fillingim, R.B.; Randich, A.; Backensto, E.M.; Faught, E. Low intensity vagal nerve stimulation lowers human thermal pain thresholds. *Pain* **2000**, *86*, 81–85. [CrossRef]

106. Sedan, O.; Sprecher, E.; Yarnitsky, D. Vagal stomach afferents inhibit somatic pain perception. *Pain* **2005**, *113*, 354–359. [CrossRef] [PubMed]
107. Ben-Menachem, E.; Revesz, D.; Simon, B.J.; Silberstein, S. Surgically implanted and non-invasive vagus nerve stimulation: A review of efficacy, safety and tolerability. *Eur. J. Neurol.* **2015**, *22*, 1260–1268. [CrossRef] [PubMed]
108. Deer, T.R.; Levy, R.M.; Rosenfeld, E.L. Prospective clinical study of a new implantable peripheral nerve stimulation device to treat chronic pain. *Clin. J. Pain* **2010**, *26*, 359–372. [CrossRef] [PubMed]
109. Deer, T.; Pope, J.; Benyamin, R.; Vallejo, R.; Friedman, A.; Caraway, D.; Staats, P.; Grigsby, E.; Porter McRoberts, W.; McJunkin, T.; et al. Prospective, Multicenter, Randomized, Double-Blinded, Partial Crossover Study to Assess the Safety and Efficacy of the Novel Neuromodulation System in the Treatment of Patients With Chronic Pain of Peripheral Nerve Origin. *Neuromodulation* **2016**, *19*, 91–100. [CrossRef] [PubMed]
110. Lee, P.B.; Horazeck, C.; Nahm, F.S.; Huh, B.K. Peripheral Nerve Stimulation for the Treatment of Chronic Intractable Headaches: Long-term Efficacy and Safety Study. *Pain Phys.* **2015**, *18*, 505–516.

Review

The Use of Neuromodulation for Symptom Management

Sarah Marie Farrell [1], Alexander Green [2] and Tipu Aziz [1,2,*]

[1]   Nuffield Department of Surgical Sciences, John Radcliffe Hospital, University of Oxford, Oxford OX3 9DU, UK; sarah.farrell@medsci.ox.ac.uk

[2]   Nuffield department of clinical Neurosciences, John Radcliffe Hospital, University of Oxford, Oxford OX3 9DU, UK; alex.green@nds.ox.ac.uk

*   Correspondence: tipu.aziz@nds.ox.ac.uk; Tel.: +44-18-6522-7645

Received: 14 August 2019; Accepted: 9 September 2019; Published: 12 September 2019

Abstract: Pain and other symptoms of autonomic dysregulation such as hypertension, dyspnoea and bladder instability can lead to intractable suffering. Incorporation of neuromodulation into symptom management, including palliative care treatment protocols, is becoming a viable option scientifically, ethically, and economically in order to relieve suffering. It provides further opportunity for symptom control that cannot otherwise be provided by pharmacology and other conventional methods.

Keywords: neuromodulation; deep brain stimulation (DBS); pain; dyspnoea; blood pressure; hypertension; orthostatic hypotension; micturition; bladder control

---

1. Introduction

Symptom management is an opportunity to alleviate suffering, whether or not a disease state is curative. This can range from an able-bodied young individual with intractable cluster headaches, to an elderly patient with non-curative cancer whose main hindrance to quality of life in their last months or years is breathlessness. The realms of palliative care extend outside that of end-of-life to encompass those living with intractable suffering. The Centre to Advance Palliative Care defines Palliative medicine as "specialized medical care for people living with serious illness" [1]. This suffering often entails symptoms of chronic pain and a variety of dysautonomias including dyspnoea, micturition dysfunction, and cardiovascular problems. For those suffering despite their current medication regime, the notion of a life free from constant pain, relief from the sensation of breathlessness, with the ability to control bladder issues, seems nothing short of a miracle.

Neuromodulation (deep brain stimulation, motor cortex stimulation, spinal cord stimulation, dorsal root ganglion stimulation) is a safe and effective treatment, largely deployed for movement disorders including Parkinson's disease tremor and dystonia [2,3], as well as epilepsy [4], psychiatric disorders such as depression/obsessive compulsive disorder/Tourette's [5,6], and a variety of previously intractable chronic pain syndromes [7,8]. Through the serendipitous amelioration of autonomic problems in patients previously fitted with these electrodes, there is a growing body of evidence demonstrating the ability of neuromodulation to ameliorate adverse autonomic effects associated with breathlessness, micturition, and cardiovascular function. These findings have led to further investigation surrounding neuromodulation and autonomic function, with kind involvement from patients previously fitted with these devices. Gaining control of these dysautonomias could palliate thousands of patients who are suffering, whether this be end-of-life or otherwise.

The brain works via a complex flow of signal processing and, when wiring is faulty, can prove difficult to correct with current mainstream pharmacology. Persuading these suboptimal networks to behave optimally is the realm of 'neuromodulation'. For the most part, this means electrodes are implanted into the brain or spine and the signals optimized to achieve a particular response, either by effectively ablating the area, or stimulating/exciting the area, though the exact mechanism of action is equivocal. The use of neuromodulation is attractive because of its reversibility, its targeted localised delivery, and the ability to adjust settings to optimize the effects (frequency, amplitude, pulse width of current delivered).

For certain conditions such as Parkinson's disease, dystonia, and tremor, neuromodulation with medical management is now established therapies and is managed by specialist neurological centres. This discussion will centre around indications for which there are little or no non-surgical alternatives. The article moves through the evidence pertaining to neuromodulation and its ability to ameliorate the symptoms of pain, hypertension, orthostatic hypotension, dyspnea, and micturition, considering the economic and ethical aspects of this care. It demonstrates the promise of neuromodulating symptoms previously intractable to pharmacology and more conventional surgery.

Methodology

A PubMed search of literature was conducted describing 'DBS' (deep brain stimulation), 'neuromodulation', or 'spinal cord stimulation' with the following search terms: 'pain' and 'autonomic function', 'cardiovascular', 'blood pressure', 'heart rate', 'micturition', 'bladder', 'respiratory', and/or 'breathing'. All references found were scanned for relevance, categorized by intractable disease type, and then reviewed in more detail. The relevant references found in these articles were also added to the list.

2. Pain

A significant number of patients suffer from intractable pain, as much as 29% of the adult population in Europe have moderate-to-severe pain [9], as well as 100 million people in the Unites States [10,11]. This carries emotional and cognitive sequelae for those it affects [12,13]. Moreover, the opioid epidemic compounds and conflates this issue. Alternatives are sought. Neuromodulation for pain is well established. Over the past few decades, the periaqueductal grey (PAG), thalamus, and more recently the anterior cingulate cortex (ACC) have become popular targets. Meanwhile, the rise of spinal cord stimulation (SCS) and dorsal root ganglion stimulation has provided a viable option.

A comprehensive review of pain and neuromodulation can be found elsewhere [7], aside from the topic of cluster headaches, which can be found below. In summary, neuromodulation is already a well-established treatment for pain, with many success stories. Outcomes are generally favourable for SCS, particularly with newer generations of technology such as burst stimulation and dorsal root ganglion implants. Studies of DBS (targeting PAG and ACC) outcomes tend to be more heterogeneous, though the Oxford group have found this treatment to be beneficial for many patients of varying pain aetiologies [14]. It might be reasonable to think that types of chronic pain, 'nociceptive versus deafferentation' or 'central versus peripheral', could be categorized as more or less amenable to neuromodulation. However, the involvement of neuronal plasticity encompasses centrally mediated changes, as shown by functional neurology and electrophysiological studies [15–17]. It may be better to select those eligible for neuromodulation by an absence of psychogenic elements, which would otherwise rule out the ability of neuromodulation to help [7]. There is a multitude of issues plaguing chronic pain trials that may lead to less-than-favourable results, including the nonrandomised nature, heterogeneity of patient aetiology, subjective unblinded assessment of patient outcome, inconsistent stimulation parameters, sites and number of electrodes implanted, the use of '50% reduction in pain' as a set threshold, and the treatment often taking place only when SCS has been unsuccessful. Thus, there are many patients for whom DBS has reduced pain and improved quality of life, but are represented as a 'fail' in the literature. The modest literature results reflect the inability of the data to represent the potential for DBS. Success rates vary between 30% and 100% pain relief in certain cases (pending long term follow up required). While it is true that published studies demonstrate an average 30% relief [18], this is no small fete for those individuals with a 30% (or more) relief in pain. Fundamentally, and of particular importance when discussing palliative treatment, for an individual with intractable pain who has already tried the limited treatment options available to them, a success rate lower than an arbitrary 50% cut off point versus a 1:500 risk of stroke from surgery may be a reasonable risk for this individual to take. Of course, the results vary depending on indications, follow-up times, dedication to optimisation of settings, and so on. Patient selection is vital, presurgical neuropsychological evaluation is required to restrict surgery to those without risk factors for negative outcomes including catastrophization, opiate addiction, low activity levels, and even ongoing litigation [19].

Posterior Hypothalamus as a DBS Target for Cluster Headaches

DBS for cluster headaches has already experienced some success: this debilitating syndrome is known to have key elements of parasympathetic activity producing the classic ipsilateral symptoms (conjunctival injection, Horner syndrome, and lacrimation) in conjunction with intense pain [20]. The hypothalamus is thought to be the source of this autonomic dysfunction on the basis of activations demonstrated in positron emission tomography (PET) and functional magnetic resonance imaging (fMRI) studies, as well as the regularity of the daily attacks in keeping with the role of the hypothalamus in circadian rhythm [21]. Stimulation of the posterior hypothalamus has been shown to increase sympathetic activity [22], and thus is utilised as a treatment for refractory cluster headache [23–25]. Success rates from continuous stimulation have been hopeful, with one study showing that 13/16 patients experienced pain abolition or major pain reduction two years after surgery, though declining to 10 patients at the four-year follow up [26]. In a prospective, double-blind crossover study in France, 11 patients were randomized to either sham of active surgery for a month followed by a one-year open phase, using weekly attack frequency as the primary outcome. Although no difference was observed between groups in the randomized phase, the one-year open phase indicated long-term efficacy in over 50% of patients without high morbidity [27].

3. Blood Pressure Control

Neuromodulation can alter cardiovascular parameters, with blood pressure (BP) being of particular interest. This has the potential to manipulate refractory hypertension (HTN) and orthostatic hypotension. The ability of the midbrain to modulate blood pressure was described in the cat in 1935 [28]. Through intraoperative investigations on the human brain, we see the predominant area of interest is the PAG, though there are intriguing findings in the posterior hypothalamic area (PHA) and the subthalamic nucleus (STN) Table 1 summarises key studies involving DBS and blood pressure.

3.1. Refractory Hypertension and Deep Brain Stimulation

Thirty-three percent of the U.S. population are affected by HTN [29], one of the greatest risk factors for cardiovascular disease. Less than half receive appropriate BP control and 0.5% of these are refractory medical treatment [30,31]. The definition of refractory HTN involves trying at least five different medications without fully rectifying BP problems, including a mineralocorticoid receptor antagonist and a thiazide-like diuretic [32,33]. It is thought to be the result of sympathetic outflow (whereas other types of HTN may be predominantly owing to hypervolaemia). Ameliorating refractory HTN is an important issue. Patients with 'resistant' HTN (elevated BP despite three or more medications) have higher risk of cardiovascular issues including stroke, myocardial infarction, aneurysm formation, heart failure, end stage renal disease, cardiac arrhythmia, hypertensive encephalopathy, hypertensive retinopathy, glomerulosclerosis, limb loss due to arterial occlusion, and sudden death [33]. While resistant HTN is better described, cardiovascular outcomes from refractory HTN indicate that these risks are further increased [33].

Table 1. Key studies involving DBS and Blood Pressure.

Author	N	Patient type	Target	Results	Conclusion
Holmberg et al. 2005 [34]	19 DBS; 10 controls with no stimulation (optimally medicated)	PD [a]	STN [b]	After 1 year of treatment, HRV [c] and BP [d] during tilt was reduced compared to baseline; but no difference between stimulated and non-stimulated group.	No beneficial results for orthostatic hypotension.
Green et al. 2005 [35]	15	PAG Chronic pain	PAG [e]	In patients at rest in a seated position, stimulation using dorsally located PAG electrodes produced an elevation of approximately 16mmHg in systolic BP, whereas stimulation using ventrally located PAG electrodes caused a decrease of approximately 14mmHg in this parameter.	PAG stimulation alters BP and is dependent on whether stimulation is dorsal or ventral.
Lipp et al. 2005 [19]	5	PD	STN	Electrodes inadvertently placed close to posterior hypothalamus showed BP and respiratory rate increased after stimulation.	PHA [f] stimulation may increase BP.
Green et al. 2006 [36]	11	PAG Chronic pain	PAG	Patients experienced decreased systolic BP in 'on' stimulation when moving from sitting to standing; in one patient with clinical orthostatic hypotension, systolic BP fell by 15% from baseline (145–148 mmHg) on changing from a sitting to standing position without stimulation, compared with a change of only 0.1% on stimulation. For those with mild orthostatic hypotension the effects were reversed. For those with orthostatic hypotension, no side effects were experienced.	Stimulation of the PAG can prevent orthostatic hypotension.
Stemper et al. 2006 [37]	14	PD	STN	DBS STN, increased HR, decreased blood flow to skin, and maintained BP after 60 degrees HUTT.	Beneficial results for orthostatic hypotension.
Green et al. 2007 [38]	1	chronic pain (oral cavity/soft palate pain)	PAG	Hypertensive patient with PAG stimulation for Chronic pain experienced their baseline a fall in arterial pressure.	PAG may be suitable as a HTN treatment.
Ludwig et al. 2007 [39]	29 (14 DBS; 15 Controls	PD	STN	BP (and HR) during rest and orthostatic conditions did not differ significantly between groups.	No beneficial results of DBS for orthostatic hypotension.
Cortelli et al. 2007 [40]	8	Cluster headache	PHA	During HUTT [g] systolic BP maintained when 'on' stimulus but fell by 3% when 'off'. Ratio of low:high frequency components in HRV increased during on stimulation.	PHA stimulation could aid orthostatic hypotension. CV (including diastolic BP) changes appear to be hypothalamic-mediated sympathetic activation.
Patel et al. 2011 [41]	1	central pain syndrome- left hemibody pain	PAG	Despite pain returning to baseline four months after surgery, DBS continued to affect BP as indicated by blood pressure rise of 18/5 mmHg.	Despite pain returning to baseline four months after surgery, DBS continued to affect BP thus affect of DBS on BP is not just relating to pain relief.
Sverrisdottir et al. 2014 [42]	17 (7 PAG, 10 STN)	chronic pain/ PD	STN	Increase in orthostatic tolerance.	Beneficial results for orthostatic hypotension.
O'Callaghan et al. 2017 [43]	1	chronic pain	ventral PAG	After 6 months of chronic low frequency DBS of vPAG, BP lowered from 280 to 210–230 systolic.	Possible use of PAG as therapy for intractable HTN.

[a] PD = Parkinson's Disease; [b] STN = Subthalamic nucleus; [c] HRV = Heart rate variability; [d] BP = Blood Pressure; [e] PAG = Periaqueductal gray; [f] PHA = Posterior hypothalamic area; [g] HUTT = Head uptilt table test.

3.2. PAG as a DBS Target for Hypertension

PAG stimulation is known to modulate cardiovascular parameters. Animal studies show electrical stimulation of the ventrolateral columns of the PAG lowers HR and BP, as well as producing freezing behaviour. In contrast, stimulation of the dorsolateral and dorsomedial produces increased heart rate and BP [35,44–46]. The separation of the PAG into four functional columns could potentially be used for differing palliative treatments. Ventrolateral stimulation would be desirable for HTN, whereas stimulation of dorsomedial and dorsolateral columns would prove useful in cases of orthostatic hypotension [38,43].

DBS PAG has produced evidence of changes in blood pressure. Green et al. found stimulation of the ventral PAG caused a 14.2 mmHg ± 3.6 mean reduction (from 143 to 128 mmHg) in systolic and 4.9 mmHg reduction in diastolic BP [47]. In line with this, one hypertensive patient—with a previously fitted PAG stimulation for chronic pain—experienced their baseline arterial pressure fall from 157.4/87.6 mmHg to 132.4/79.2 mmHg [41]. More recently, O'Callaghan et al. reported a patient who had previously tried a variety of medications to control her blood pressure, including a baroreceptor activation device. After six months of ventral PAG stimulation, her average morning blood pressure dropped from a pre-surgery value of 280/166 mmHg to 210/130 mmHg [48].

Similar blood pressure reductions can be sustained one year after surgery [36,49,50]. While it is possible these reductions may be confounded by analgesic benefit, one study demonstrated a separation between the two. Left hemibody pain (under PAG stimulation) returned to baseline 4 months after surgery, but at 27 months, when DBS stimulation was turned off, blood pressure rose by 18.5 mmHg, indicating that there had been a continued effect separate from pain relief [17].

3.3. PAG as a DBS Target for Orthostatic Hypotension

The original Green et al. study from 2005, demonstrating a decrease in BP upon ventral stimulation, also showed dorsal stimulation to have the opposite effect, that is, mean increase in systolic BP of 16.7 ± 5.9 mmHg (n = 6) [47]. A further study by the same group explored whether orthostatic hypotension could be treated with PAG stimulation. Eleven patients previously implanted for chronic pain were divided into three groups—those with orthostatic hypotension (n = 1), those with mild orthostatic intolerance (n = 5), or those with no orthostatic intolerance (n = 5). When 'on' stimulation, the decrease in systolic BP while moving from sitting to standing was reduced from 28.2% to 11.1% in the patient with orthostatic hypotension [51]. For those with mild orthostatic intolerance, the systolic BP (initially 15.4% drop) was completely reversed. The remaining group experienced no side effects. This amelioration of orthostatic hypotension was found without increasing baseline resting BP, but with an associated increased heart rate response to standing.

The effects of PAG stimulation on the cardiovascular system may be understood as there is evidence that PAG projects to preganglionic cardiac vagal neurons in the nucleus ambiguous, with chemical stimulation of PAG in rats inhibiting baroreflex vagal bradycardia [34].

Overall, ventral PAG stimulation seems beneficial for HTN, whereas dorsal PAG stimulation may improve orthostatic hypotension. Given that orthostatic hypotension is present in 30% of adults 70 years and older [39], contributing to increased fall risk, cardiovascular disease, stroke, and death [37], the ability to mitigate these BP difficulties is clinically significant. We have yet to purposefully target ventral versus dorsal columns, though there is no reason this could not be attempted in future. Given this convincing evidence of the effect of PAG on BP, it is surprising to note there are no clinical trials registered at clinicaltrials.gov (using search terms of 'DBS', 'neuromodulation', 'deep brain stimulation' and 'blood pressure', 'hypertension', or 'orthostatic hypotension'). This may speak to the low number of PAG stimulated chronic pain patients available for the research.

3.4. Subthalamic Nucleus as a DBS Target for Orthostatic Hypotension

Although two studies of head up tilt table testing (HUTT) in patients undergoing DBS of the STN reported no beneficial results [42,52], several studies demonstrate an increase in, or maintenance of, both arterial BP and baroreceptor sensitivity [53–55]. Stemper et al. demonstrated that patients with STN DBS were able to maintain BP and baroreflex sensitivity, without which they experienced significant orthostatic hypotension during HUTT [53]. As neither BP or baroreflex sensitivity were influenced by stimulation of motor thalamus, globus pallidus interna (GPi), pedunculopontine nucleus (PPN), or ACC, this suggests some anatomic specificity to STN. In keeping with this, Hyam et al. demonstrated that PD patients who are stimulated in the STN with a frequency over 100 Hz experience a modest increase in HR and BP (5 bpm and 5 mmHg, respectively) [36]. Sverrisdottir et al. also demonstrated increase in orthostatic tolerance in Parkinson's disease (PD) patient with STN DBS [54].

Although some studies show this improvement in orthostatic hypotension is sympathetically-mediated [40,56,57], Sumi et al. highlighted the possibility that any cardiovascular improvements seen with STN DBS could be the result of the increased ability to exercise rather than the stimulation itself [58], or a reduced pharmacological requirement of the Parkinson's disease state [58]. However, peri-operative studies showing cardiovascular changes suggest there may be a more direct role of STN. Hyam points out this may be because of the high frequencies of STN leading to exposure of stimulus to nearby areas of the central autonomic networks [36]. This is supported by earlier findings that showed the one STN patient (out of the five) who experienced autonomic alterations was the same patient that had a lead placement extending to posteromedial and lateral hypothalamic areas [22].

3.5. Posterior Hypothalamus as a DBS Target to Ameliorate Orthostatic Hypotension

With the advent of using DBS for PD, multiple patients have been tried and tested, allowing variation across electrode placement positions. Electrodes inadvertently placed close to posterior hypothalamus showed that BP and respiratory rate increased after stimulation [22]. It would seem that stimulation of posterior hypothalamus may facilitate the maintenance of SBP during head up-tilt testing. As previously mentioned, patients with cluster headaches can have stimulation in the posterior hypothalamic area (HPA). When off stimulation, HUTT resulted in a 3% fall in systolic blood pressure, whereas systolic BP was maintained when the test was repeated during stimulation [22].

4. Dyspnoea in Chronic Obstructive Airway Disease

Dyspnoea is difficult to treat and distressing for the patient. It is the most important complaint in many common respiratory diseases such as chronic obstructive airway disease (COPD) [59]. COPD affects 6% of the population and is a major cause of morbidity and mortality worldwide [60]. Many other diseases lead to this distressing sensation of breathlessness including asthma, lung cancer, and end-stage heart failure, to name a few. Neuromodulation may be able to tap into the central control signals and provide some relief. Bronchoconstriction causes increased airway resistance and is a key element of the pathology behind asthma and COPD. Airway smooth muscle is mediated by airway-related vagal pre-ganglionic cells (AVPN)—part of the parasympathetic nervous system. Relaxing these airways requires the circulating catecholamines of the sympathetic nervous system. It is, therefore, feasible to control the lungs via the brain.

Since 1894, the effect of orbitofrontal cortex stimulation on respiratory rate was demonstrated in a variety of animals including cats, dogs, and monkeys [61]. Manipulation of the respiratory rate was later shown by stimulation of the anterior cingulate cortex during open brain surgery [62].

Hyam et al. studied the effect of stimulation on peak expiratory flow rate (PEFR) and forced expiratory volume in one second (FEV1) in 17 chronic pain patients and 20 movement disorder patients. Within the pain syndrome patients, 10 had stimulation in the PAG (an area with connections to AVPN) and 7 in the sensory thalamus. A similar control was used for the movement disorders, with 10 STN stimulated patients (relevant to the AVPN) and 10 GPi stimulation patients. Spirometry recordings were made and the PEFR was taken, along with the FEV1. PEFR is defined as the highest flow achieved from a maximum forced expiratory manoeuvre started without hesitation from a position of maximal lung inflation. FEV1 is defined as the maximal volume of air exhaled in the first second of a forced expiration from a position of full inspiration. The experimenters also recorded the maximal volume of air exhaled with maximally forced effort from maximal inspiration, or forced vital capacity (FVC).

STN and PAG both improved PEFR, but not FEV1. Mean PEFR percentage change was 13.4 ± 4.6% with PAG stimulation, a non-significant 0.89 ± 2.6% with sensory thalamus stimulation, 14.5 ± 5.3% with STN stimulation, and −0.2 ± 1.8% with GPi stimulation [63]. The lack of significant change in FEV1 could be related to either the STN and PAG affecting upper airways more, or the ability of airway flow changes in FEV1 to be seen only if the patients suffered from respiratory disease. This is bolstered by the one patient who showed an obstructive lung function profile on spirometry—exhibiting a 9.8% increase in FEV1 on PAG stimulation. This effect of stimulation is comparable to the effects of nebulised or oral steroid use seen in acute exacerbation of COPD [64–66] and lacks the risk of toxic side effects. Given that patients stimulated in the sensory thalamus and GPi showed no change in lung function, we can assume these changes are not related to the decrease in pain or amelioration of motor disorder.

It is reasonable to suppose that STN stimulation improves lung function, having previously been implicated in the respiratory network. STN is expected to be active during breath holding and has a role in inhibiting initiated responses in stop-signal paradigms [67,68]. The PAG is also recognised to be integral in the freeze–flight phenomenon, hence stimulation of PAG causes changes in the cardiovascular system, vocalization, and micturition [69–71]. It is logical that these sympathetically innervated actions would also extend to increasing respiratory function, preparing the body to fight or flee. The PAG projects to the parabrachial nuclei and stimulation of the nuclei in animals have been implicated in cardiorespiratory variables: lesions in this region can cause distortions of the Hering–Breuer reflex [72,73], whereas chemical stimulation in anaesthetized cats causes reduced total lung resistance [74,75].

The management of dyspnoea in patients with chronic airway diseases requires more attention; currently, the American Thoracic Society recommends bronchodilators as the mainstay of treatment, with lung volume reduction surgery and anxiolytic therapies considered on an individual basis [76]. It would be extremely interesting to see the benefit of DBS in COPD, asthma, and sleep apnoea patients. To our knowledge, there is currently only one clinical study registered on clinicaltrials.gov related to DBS and respiratory function. This study, among other things, will look at how 'on' versus 'off' stimulation affects patients with multisystem atrophy who tend to experience decreased respiratory function. See Table 2 for clinical trials relating to pain and dysautonomias and Table 3 for key studies surrounding DBS and dyspnoea.

Table 2. Unpublished Registered Clinical Trials relating to DBS and Palliative symptoms.

Area	Title	Status	Conditions	Interventions	Locations
Blood Pressure	Deep Brain Stimulation for Autonomic and Gait Symptoms in Multiple System Atrophy	Recruiting	Multiple System Atrophy|Autonomic Failure|Postural Hypotension|Bladder, Neurogenic|Gait Disorders, Neurologic	Procedure: Deep brain stimulation	John Radcliffe Hospital, Oxford, Oxfordshire, United Kingdom
Chronic pain	Deep Brain Stimulation (DBS) for Chronic Neuropathic Pain	Recruiting	Chronic Neuropathic Pain|Post Stroke Pain|Phantom Limb Pain|Spinal Cord Injuries	Device: Active DBS|Device: Inactive DBS	University of California, San Francisco, California, United States
Chronic pain	Safety Study of Deep Brain Stimulation to Manage Thalamic Pain Syndrome	Completed	Chronic Pain	Device: Deep Brain Stimulation for Thalamic Pain Syndrome	Cleveland Clinic Foundation, Cleveland, Ohio, United States
Chronic pain	Combined Cingulate and Thalamic DBS for Chronic Refractory Chronic Pain	Not yet recruiting	Chronic Refractory Neuropathic Pain	Procedure: Deep brain Stimulation of cingulum anterior	Department of neurosurgery, Nice, France
Respiratory dysfunction	The Effects Of DBS Of Subthalamic Nucleus On Functionality In Patients With Parkinson's Disease: Short-Term Results	Recruiting	Parkinson Disease|Surgery|Respiration; Decreased|Muscle Weakness	Device: Maximum Inspiratory Pressure and Maximum Expiratory Pressure	Hatay Mustafa Kemal University, Antakya, Hatay, Turkey
Urinary dysfunction	Effect of Deep Brain Stimulation on Lower Urinary Tract Function	Completed	Movement Disorder|Urinary Tract Disease	Procedure: deep brain stimulation ON|Procedure: Deep brain stimulation OFF	Department of Neurology, University of Bern, Bern, Switzerland|Department of Urology, University of Bern, Bern, Switzerland
Urinary dysfunction	Deep Brain Stimulation in Patients With LUTS	Recruiting	Bladder Dysfunction|Neurogenic Bladder	Other: Cohort	Houston Methodist Research Institute, Houston, Texas, United States
Other dysfunction	Deep Brain Stimulation and Digestive Symptomatology	Completed	Parkinson's Disease	not specified	Rouen University Hospital, Rouen, France

Table 3. Key Studies surrounding DBS and Dyspnoea.

Author	N	Patient Disease	Target	Outcome Measures	Results	Conclusion
Hyam et al. 2012 [61]	17 chronic pain; 20 movement disorder; 7 control thalamus; 10 control Gpi	Movement disorder and chronic pain	STN [a] and PAG [b] (sensory thalamus and GPi [c] as control)	PEFR [d],FEV1 [e]	STN and PAG improved PEFR but not FEV1; Patient with obstructive lung function showed improved FEV1 on stimulation.	Possible to control the lungs via the brain.
Vigneri et al. 2012 [76]	6 PD; 5 cluster headache	PD [f] and cluster headaches	STN and PHA [g]	RR [h] (and HR [i], BP [j])	No change on vs. off stimulation for any values when supine.	Failure to find effect.
Xie et al. 2015 [77]	1	PD	STN	Respiratory dyskinesia	Stimulation relieved levodopa-induced respiratory distress even in medication 'on' phase.	Not clear how DBS controls respiratory dyskinesia.

[a] STN = subthalamic nucleus; [b] PAG = Periaqueductal gray; [c] GPi = globus pallidus; [d] PEFR = Peak expiratory flow rate; [e] FEV1 = Forced Expiratory Volume in 1 second; [f] PD = Parkinson's disease; [g] PHA = posterior hypothalamic area; [h] RR = respiratory rate; [i] HR = heart rate; [j] BP = blood pressure.

5. Micturition

Lower urinary dysfunction (LUTD) is disabling [77], difficult to treat, and extremely common in neurological diseases such as PD [78–80] and multiple sclerosis [81]. In fact, these symptoms occur in 74% of patients with PD, with severe symptoms in over 50% [82].

LUTD is an increasing common symptom worldwide referring to urgency, increased urinary frequency, or incontinence, with neurological disorder being one of the main causes. Urodynamic examination in these patients shows detrusor hyperreflexia with involuntary contractions of the bladder, resulting in a reduced bladder capacity and an early desire to void. Sacral neuromodulation (SNM) is established as a treatment for non-neurogenic LUTD, but a more centralised method may be required to achieve results for neurogenic LUTD given the lack of widespread success of SNM for LUTD [83]. It is feasible that central neuromodulations are a potential treatment approach given the complex network of autonomic and central nervous system control of micturition reflexes [84,85]. The pontine micturition centre (PMC) is the central micturition reflex centre for the brain [86,87]. Pathways of the lower urinary tract system consist of an afferent pathway from the urinary bladder to the PMC via the pelvic nerve and spinal cord, and an efferent pathway projecting from the PMC to the bladder via the sacral parasympathetic centre of intermediolateral column cells [88]. The PAG, locus coeruleus (LC), and rostral pontine reticular nucleus are all thought to modulate bladder activity via their connections to the PMC.

We can think of the bladder as being in a 'storage mode' and switching to voiding when socially appropriate. Research looking to optimise urinary function using DBS stems from movement disorder and chronic pain patients willing to undergo urodynamic testing.

5.1. STN and Micturition

Both animal and human studies show improved urodynamics with STN stimulation, measured by bladder capacity and first desire to void [89–92]. For example, Seif et al. report initial desire to void at 199 ± 57 mL, with maximal bladder capacity at 302 ± 101 mL. This contrasts with the 'off' stimulation condition (initial desire to void occurring at mean value of 135 ± 43 mL). The maximal capacity of the bladder was 174 ± 52 mL, with both values shown to be significantly different in the on and off stimulation ($p < 0.005$) [92]. Thus, DBS of STN reduces detrusor overactivity and increases bladder capacity, effectively normalising the 'storage' phase.

Positron emission tomography (PET) studies show that 'off' stimulation bladder-filling in PD patients with bilateral STN stimulation increases regional cerebral blood flow to ACC and lateral frontal cortex [91]. This suggests that 'on' stimulation may ameliorate the bladder dysfunction by effective integration of afferent bladder information. Future studies should address the effects of STN DBS in patients specifically with continence problems.

5.2. Thalamus and Micturition

Studies of thalamus and micturition suggest a negative effect on lower urinary tract symptoms. Thalamic DBS has been shown to significantly decrease bladder volume at 'first' and 'strong' desire to void, as well as decreasing maximal bladder capacity [93,94]. This demonstrates that although the thalamus has a role in micturition, it is not a target for rectifying dysfunction.

5.3. PAG and Micturition

The PAG has previously been hypothesized to be a micturition switch, changing bladder state from 'storage' to 'voiding'. It is responsible for sensory inputs from distended bladder-activated spinal–midbrain–spinal nerve circuitry. Lumbrosacral neurons are known to terminate on neurons in the PAG [95], and the PAG then projects to the PMC [96]. PAG neurons are activated during voiding [97], and animal studies have previously shown that PAG elicits micturition effects—both stimulatory [95,98] and inhibitory [99].

In humans, stimulation in PAG increases maximal bladder capacity, as judged by the volume at which patients, fitted with a catheter, ask for a saline infusion to be stopped. Increased subjective bladder capacity was found in the 'on' versus 'off' state ($p = 0.028$), though this did not affect volumes at which voiding was desired [93]. Of note, when 'on' stimulation while the bladder was filling from empty, the volume at which urine first started to escape from the penis (maximum cystometric capacity (MCC)) was greater compared with the 'off' stimulation. Interestingly, stimulation affected micturition over a much wider area of the PAG than the expected caudal ventrolateral part [100]. The Oxford group hypothesizes that rostrally located electrode placements in the PAG are most likely to activate afferent inputs to the caudal ventrolateral PAG [101], whereas more caudal stimulation activates local intrinsic connections to the ventrolateral PAG [100] or possibly descending efferent pathways [102,103]. It is also possible that the afferent pathway has antidromic activity from the cord, serving to cancel orthodromic afferent signals from the bladder. Activation of any of these circuits by DBS could potentially disrupt the micturition control network in the PAG either via the GABAergic synapses or by creating a non-physiological network activity pattern, blocking initiation of a synchronized voiding pattern in the detrusor and sphincter muscles. Thus, urine output is prevented even when the bladder is distended [97].

5.4. PPN and Micturition

The pedunculopontine nucleus (PPN) is a relatively new experimental brain target for managing severe Parkinson's disease, and through studies relating to this, its involvement in micturition is highlighted. PPN DBS is known to improve akinesia and gait difficulties in both animals and patients with PD [104–108]. In 2011, Aviles-Olmos et al. described their findings of this stimulation causing detrusor over-activity immediately after right-sided PPN DBS [104]. This is supported by a study of Gottingen minipigs, where stimulation of the PMC resulted in increased detrusor pressure [109], measured by cystometry, and is plausibly because of the involvement of the pontine micturition centres and their connections.

Despite this, the Oxford group did not find detrusor overactivity of lower sensory threshold during bladder filling. They studied five patients with bilateral PPN for PD. In fact, stimulation provided significant increase in maximal bladder capacity averaging at 199 mL during the 'on' stimulation (range 103–440 mL) compared with 131 mL during the 'off' stimulation (range 39–230 mL) [110]. It is worth noting the considerable spread of response to stimulation in terms of bladder capacity, across subjects. Further investigation is required to attribute this to any particular cause, given that it was not linked to stimulation type (monopolar vs. bipolar), electrode location, or duration of stimulation. Interestingly, white matter tractography did not show either modulation of activity within the PAG (a proposed 'micturition switch') or involvement of many established bladder network components (e.g., insula cortex and ACC). An improved understanding of what is causing these varied and differing results are required before trials of this implant are made purely for micturition issues alone.

In summary, the growing literature of DBS and micturition provides potential targets. Basal ganglia and brainstem targets (STN and PAG) inhibit micturition and improve incontinence. The results from the PPN also appear promising. However, thalamic targets induce micturition [81]. A 2017 DBS study in rats was the first study to evaluate the effects of conducting DBS on four potential targets on bladder activities. It suggests the Pedunculopontine tegmental nucleus is the most promising DBS target for developing new approaches to treat bladder dysfunction, being most efficient in suppressing reflexive isovolumetric bladder contractions compared with PAG, rostral pontine reticular nucleus, and locus coeruleus [111].

The rapid onset and reversibility of DBS allows the prospect of intermittent, patient-controlled use with minimal risk of side effects. This may be of particular use in those urinary incontinence syndromes of central origin such as Parkinson's disease or poststroke incontinence. Current clinical trials pertaining to DBS and lower urinary tract symptoms are listed in Table 2. Table 4 lists key studies surrounding DBS and Micturition.

Table 4. Key Studies surrounding DBS and Micturition.

Author	Patient N	Patient Type-Disease And Brain Area	Target	Outcome Measure	Results	Conclusion
Finazzi-Agro, Peppe et al. 2003 [88]	5	PD [a]	STN [b] bilateral	Bladder compliance and capacity, first desire to void volume, bladder volume (reflex volume) and amplitude of detrusor hyperreflexic contractions, maximum flow, detrusor pressure at maximum flow and detrusor-sphincter coordination	Bladder capacity and reflex volume were increased for 'on' stimulation. No significant differences in other parameters.	STN stimulation seems to be effective for decreasing detrusor hyperreflexia in PD. Hence a role for basal ganglia in voiding control.
Seif et al. 2004 [90]	16	PD	STN	Filled bladder with isotonic saline and measured intiial desire to void, maximal bladder capcity, detursor contractions, detrusor pressure, and compliance of bladder.	Maximal bladder cacpity and volume at desire to void are both significantly higher during 'on' stimulation.	DBS of STN normalises 'storage' phase of micturition.
Herzog et al. 2006 [89]	11	PD	STN bilateral	PET [c] studies measuring regional cerebral blood flow to ACC [d] during bladder filling.	During bladder filling, 'off' stimulation demonstrates increased regional blood flow to ACC and lateral frontal cortex.	Stimulation aids more effective integration of afferent bladder information.
Winge and Nielson 2010 [78]	107	PD	STN	Danish Prostate Symptom Score; International prostate symptom score.	Patients with DBS had lower reporting of nocturia compared to patients with apomorphine pumps.	STN DBS results in lower levels of nocturia.
Green, Stone et al. 2012 [91]	6 PAG; 2 control thalamus	Chronic pain	PAG [e]	Sensation of filling and desire to void during saline infusion 'on' and 'off' stimulation.	On' stimulation: Volume urine first escaped from penis was higher 'on', and subjective bladder capacity was increased. This did not affect volumes at which voiding was desired.	It is possible PAG stimulation can switch off micturition.
Aviles-Olmos, Foltynie et al. 2011 [102]	1	PD	PPN [f] rt side	NA	Detrusor overactivity/urge incontinence after DBS surgery. Voiding normal. Symptoms improved 6 months post-op with antimuscarinics.	Involvement of pontine micturition centres resulted in urge uncontinence.
Roy et al. 2018 [108]	6 (5 complete)	PD	PPN bilateral	Bladder volume at maximal capacity (also looked at white matter tractography).	Increase in maximal bladder capacity when 'on' stimulation.	PPN may be a target to alleviate some LUTD symptoms.
Kessler, Burkhard et al. 2008 [92]	7	Essential Tremor	Thalamus	Bladder volume at first desire to void, maximal cystometric capacity.	Stimulation decreases bladder volume at 'first' and 'strong' desire to void, and maximal bladder capacity.	Thalamus may have a role in micturition but is not a target for rectifying dysfunction.

[a] PD = Parkinson's Disease; [b] STN = Subthalamic Nucleus; [c] PET = Positron Emission Tomography; [d] ACC = Anterior Cingulate Cortex; [e] PAG = Periqueductal Gray; [f] PPN = Pedunculopontine Nucleus.

5.5. The Potential of DBS for Other Intractable Symptoms

Neuromodulation may also prove useful for other debilitating conditions proving intractable and carrying substantial morbidity. For example, sudomotor dysfunction is extremely common in PD patients and can make for uncomfortable sleep and awkward social functioning. One study showed that 34 out of 35 patients were completely relieved of the drenching sweats they experienced prior to STN stimulation [112] and several others demonstrate similar effects [77,113,114]. In contrast, electrodes mistakenly placed in thalamus or posterolateral hypothalamus have caused hyperhidrosis [22,115].

Additionally, there is evidence for the improvement of GI dysmotility with STN stimulation, including improved deglutition and faster pharyngeal transit times, leading to a reduced rate of aspiration during swallowing [116–118]. Improved gastric emptying has been demonstrated by 13C-acetate breath testing [40,119] and improved GI motility [40]. The latter study calculated frequency reductions of 50%–25% for dysphagia, 35–15% for sialorrhea, 95–75% for constipation, and 85–50% for problems with defecation.

6. Cost Effectiveness of Neuromodulation for Intractable Suffering

There is no doubt that the list of symptoms discussed spans millions of people. Chronic pain alone affects approximately 100 million people in the United States, costing 635 billion dollars each year in medical treatment and lost productivity [11]. Hypertension effects 32.6% of the U.S. population [29], 5% of which is refractory to current medications [20,31]. Dyspnoea spans a whole host of respiratory diseases including COPD, heart failure, lung cancer, fibrosis, and asthma. Urinary urgency, frequency, nocturia, and incontinence are common symptoms in many neurological diseases including multiple sclerosis and Parkinson's disease.

The economic sequelae of palliating these symptoms in so many patients are difficult to calculate. Essentially, it is easy to see the costs, and hard to quantify benefits. There is an initial high cost of implants and follow up clinics. A systematic review of DBS for PD calculated that over five years, the cost of DBS for one patient would be 186,244 USD (amalgamated from nine studies) [120], and it would be reasonable to assume a similar value for other uses of DBS.

It is more difficult to weigh up the economic gains. These include increased productivity for those able to re-enter the workforce, less time waiting for clinic appointments or drug trials, reduced requirement of pharmacology therapy, reduced need for carers once symptoms have ameliorated, and patients can resume activities of daily living, and so on. It is worth noting that the economics of these changes depend on whether a patient is nearing end-of-life, as does their suitability on a fitness-for-surgery basis. The expansion of palliative care outside that of end-of-life shifts the balance towards neuromodulation.

7. Ethical Considerations for Symptom Management

While reduction in suffering is always desirable, deriving the extent to which finite resources should be devoted to this goal is more complex. A second layer of complexity results from the need to evaluate the extent to which additional risk should be incurred attempting interventions that may not be successful. This is particularly difficult in a palliative setting [121].

The ethics of providing neuromodulation is a complex formula calculated on a case-by-case basis including a multitude of factors: the general medical condition of the patient, effectiveness of the therapy in question, comparative effectiveness of other therapies available, amount of suffering encountered by the patient, life expectancy, and of course the cost of the therapy as it relates to the ability to provide for others with the resources available. On the latter point, it is hoped that neuromodulation can avoid increasing symptom management costs compared with current best practice, for example, by avoiding hospitalizations late in disease course. Parallels may be drawn here with chemotherapy costs in pain management, where research demonstrates that, compared with less expensive treatments, initial costs are levelled out with reductions in the care required further down the line [122].

8. Conclusions

There are millions of patients suffering from intractable debilitating symptoms whose treatment needs are not currently being met by conventional medications. Neuromodulation is a promising palliative method. Evidence weighs in heavy on the positive effect of neuromodulation for intractable pain, particularly through stimulating the PAG and ACC. This comes with the caveat of treating the right patient, in the right brain location, and with the right stimulation parameters. The patient must work with neurosurgeons and the rest of the pain team to find their 'Goldilocks' parameter settings.

Increasingly, research demonstrates a role for neuromodulation of autonomic dysfunction in a variety of settings ranging from blood pressure control, to micturition, to breathlessness. Firstly, it is theoretically possible that both uncontrolled hypertension and postural hypotension may be amenable to DBS in the future. Particular locations of interest include the PAG for HTN and STN for orthostatic hypotension. This has already been demonstrated in a small number of patients, both short and longer term. Secondly, evidence is accumulating for the control of micturition, which may help those suffering from detrusor instability and early desire to void. Future research can focus on the subgroup of patients (e.g., Parkinson's patients) suffering from this and already receiving neuromodulation for motor problems. Current trials are listed in Table 2. Locations of interest include STN and PPN, with some unresolved hints that PAG may be of interest. Thirdly, while many potential patients could benefit from neuromodulation for dyspnoea, the research is in its infancy, but highlights the importance of exploring this potential avenue in what is otherwise a neglected area of research. Successful research into the effect of PAG and STN stimulation on PEFR can now be parlayed into examining patients with dyspnoea as a predominant issue. For example, exploration of DBS in patients with abnormal lower airway calibre and established chronic lung disease to confirm the benefits of controlling the lungs via the brain, and to more fully understand the mechanisms.

Although we are yet to map out the autonomic nervous system in a way that provides full mastery over it, we have enough tantalizing leads to create targets that provide invaluable relief from a whole host of distressing symptoms, regardless of incomplete mechanistic framework. Lest we not forget, this is a compromise that has worked very well for Parkinson's and tremor patients. One of the appealing features of trialling neuromodulation in refractory states is its reversibility. This is partially because there is a sense of the ability to 'undo' should it not go to plan, and partially because this on–off ability may prove useful as part of the treatment. For example, for those with orthostatic hypotension, stimulation could be halted overnight when patients are supine, preventing the nocturnal hypertension these patients experience using current medications [123]. Additionally, neuromodulation offers the ability to change settings as symptoms change, compared with, for example, the 'one shot' opportunity of a cingulotomy for intractable cancer pain [124].

The use of neuromodulation could revolutionise symptom-control in the near future, providing services for those difficult to reach under current regimes. The challenges are to justify the initial cost of surgery, carefully select the right patients, and acquiesce neurophobia for all involved (medical team, patients, relatives, and funding bodies). We can now start to tap into the potential benefits in a manner akin to what has already developed for motor symptoms.

Author Contributions: S.M.F. performed original draft preparation; A.G., T.A., and S.M.F. performed writing (reviewing and editing).

Funding: Research for this review article received no external funding.

Acknowledgments: The research was supported by the National Institute for Health Research (NIHR) Oxford Biomedical Research Centre based at Oxford University Hospitals NHS Trust and university of Oxford.

Conflicts of Interest: The authors declare no conflict of interest.

References

1. Listed, N.A. Cancer Pain Relief and Palliative Care. Report of a WHO Expert Committee. Available online: https://apps.who.int/iris/handle/10665/39524 (accessed on 12 September 2019).
2. Crowell, J.L.; Shah, B.B. Surgery for Dystonia and Tremor. *Curr. Neurol. Neurosci. Rep.* **2016**, *16*, 22. [CrossRef] [PubMed]
3. Hickey, P.; Stacy, M. Deep Brain Stimulation: A Paradigm Shifting Approach to Treat Parkinson's Disease. *Front. Neurosci.* **2016**, *10*, 173. [CrossRef] [PubMed]
4. Salanova, V. Deep brain stimulation for epilepsy. *Epilepsy Behav.* **2018**, *88S*, 21–24. [CrossRef] [PubMed]
5. Drobisz, D.; Damborská, A. Deep brain stimulation targets for treating depression. *Behav. Brain Res.* **2019**, *359*, 266–273. [CrossRef] [PubMed]
6. Karas, P.J.; Lee, S.; Jimenez-Shahed, J.; Goodman, W.K.; Viswanathan, A.; Sheth, S.A. Deep Brain Stimulation for Obsessive Compulsive Disorder: Evolution of Surgical Stimulation Target Parallels Changing Model of Dysfunctional Brain Circuits. *Front. Neurosci.* **2018**, *12*, 998. [CrossRef] [PubMed]
7. Farrell, S.M.; Green, A.; Aziz, T. The Current State of Deep Brain Stimulation for Chronic Pain and Its Context in Other Forms of Neuromodulation. *Brain Sci.* **2018**, *8*, 158. [CrossRef]
8. Vyas, D.B.; Ho, A.L.; Dadey, D.Y.; Pendharkar, A.V.; Sussman, E.S.; Cowan, R.; Halpem, C.H. Deep Brain Stimulation for Chronic Cluster Headache: A Review. *Neuromodulation* **2018**, *22*, 388–397. [CrossRef]
9. Breivik, H.; Collett, B.; Ventafridda, V.; Cohen, R.; Gallacher, D. Survey of chronic pain in Europe: Prevalence, impact on daily life, and treatment. *Eur. J. Pain* **2006**, *10*, 287–333. [CrossRef]
10. Pizzo, P.A.; Clark, N.M.; Carter-Pokras, O. *Relieving Pain in America: A Blueprint for Transforming Prevention, Care, Education, and Research*; National Academies Press: Washington, DC, USA, 2011.
11. Sciences NAO. Relieving Pain in America: A Blueprint for Transforming Prevention, Care, Education, and Research. *Mil Med.* **2016**, *181*, 397–399. [CrossRef]
12. Elliott, A.M.; Smith, B.H.; Penny, K.I.; Smith, W.C.; Chambers, W.A. The epidemiology of chronic pain in the community. *Lancet* **1999**, *354*, 1248–1252. [CrossRef]
13. McCracken, L.M.; Iverson, G.L. Predicting complaints of impaired cognitive functioning in patients with chronic pain. *J. Pain Symptom Manag.* **2001**, *21*, 392–396. [CrossRef]
14. Boccard, S.G.J.; Prangnell, S.J.; Pycroft, L.; Cheeran, B.; Moir, L.; Pereira, E.A.C.; Fitzgerald, J.J.; Green, A.L.; Aziz, T.Z. Long-Term Results of Deep Brain Stimulation of the Anterior Cingulate Cortex for Neuropathic Pain. *World Neurosurg.* **2017**, *106*, 625–637. [CrossRef]
15. Anderson, W.S.; O'Hara, S.; Lawson, H.C.; Treede, R.D.; Lenz, F.A. Plasticity of pain-related neuronal activity in the human thalamus. *Prog. Brain Res.* **2006**, *157*, 353–364. [CrossRef]
16. Coderre, T.J.; Katz, J.; Vaccarino, A.L.; Melzack, R. Contribution of central neuroplasticity to pathological pain: Review of clinical and experimental evidence. *Pain* **1993**, *52*, 259–285. [CrossRef]
17. Schweinhardt, P.; Lee, M.; Tracey, I. Imaging pain in patients: Is it meaningful? *Curr. Opin. Neurol.* **2006**, *19*, 392–400. [CrossRef]
18. Frizon, L.A.; Yamamoto, E.A.; Nagel, S.J.; Simonson, M.T.; Hogue, O.; Machado, A.G. Deep Brain Stimulation for Pain in the Modern Era: A Systematic Review. *Neurosurgery* **2019**. [CrossRef]
19. Mole, J.A.; Prangnell, S.J. Role of clinical neuropsychology in deep brain stimulation: Review of the literature and considerations for clinicians. *Appl. Neuropsychol. Adult* **2017**, *26*, 283–296. [CrossRef]
20. van Viljet, J.A.; Vein, A.A.; Ferrari, M.D.; Van Dijk, J.G. Cardiovascular autonomic function tests in cluster headache. *Cephalgia* **2006**, *26*, 329–331.
21. May, A.; Ashburner, J.; Büchel, C.; McGonigle, D.J.; Friston, K.J.; Frackowiak, R.S.; Goadsby, P.J. Correlation between structural and functional changes in brain in an idiopathic headache syndrome. *Nat. Med.* **1999**, *5*, 836–838. [CrossRef]
22. Lipp, A.; Tank, J.; Trottenberg, T.; Kupsch, A.; Arnold, G.; Jordan, J. Sympathetic activation due to deep brain stimulation in the region of the STN. *Neurology* **2005**, *65*, 774–775. [CrossRef]
23. Leone, M. Deep brain stimulation in headache. *Lancet Neurol.* **2006**, *5*, 873–877. [CrossRef]
24. Leone, M.; Franzini, A.; Bussone, G. Stereotactic stimulation of posterior hypothalamic gray matter in a patient with intractable cluster headache. *N. Engl. J. Med.* **2001**, *345*, 1428–1429. [CrossRef]
25. Leone, M.; Proietti Cecchini, A. Deep brain stimulation in headache. *Cephalalgia* **2016**, *36*, 1143–1148. [CrossRef]

26. Leone, M.; Franzini, A.; Broggi, G.; Bussone, G. Hypothalamic stimulation for intractable cluster headache: Long-term experience. *Neurology* **2006**, *67*, 150–152. [CrossRef]

27. Fontaine, D.; Lazorthes, Y.; Mertens, P.; Blond, S.; Geraud, G.; Fabre, N.; Navez, M.; Lucas, C.; Dubois, F.; Gonfrier, S. Safety and efficacy of deep brain stimulation in refractory cluster headache: A randomized placebo-controlled double-blind trial followed by a 1-year open extension. *J. Headache Pain* **2010**, *11*, 23–31. [CrossRef]

28. Kabat, H.; Magoun, H.W.; Ranson, S.W. Electrical stimulation of points in the forebrain and midbrain. The resultant alteration in blood pressure. *Arch. Neurol. Psychiatry* **1935**, *34*, 931–955. [CrossRef]

29. Elijovich, F.; Weinberger, M.H.; Anderson, C.A.; Appel, L.J.; Bursztyn, M.; Cook, N.R.; Dart, R.A.; Newton-Cheh, C.H.; Sacks, F.M.; Laffer, C.L. Salt Sensitivity of Blood Pressure: A Scientific Statement from the American Heart Association. *Hypertension* **2016**, *68*, e7–e46. [CrossRef]

30. Calhoun, D.A.; Booth, J.N.; Oparil, S.; Irvin, M.R.; Shimbo, D.; Lackland, D.T.; Howard, G.; Safford, M.M.; Muntner, P. Refractory hypertension: Determination of prevalence, risk factors, and comorbidities in a large, population-based cohort. *Hypertension* **2014**, *63*, 451–458. [CrossRef]

31. Howard, G.; Prineas, R.; Moy, C.; Cushman, M.; Kellum, M.; Temple, E.; Graham, A.; Howard, V. Racial and geographic differences in awareness, treatment, and control of hypertension: The Reasons for Geographic and Racial Differences in Stroke study. *Stroke* **2006**, *37*, 1171–1178. [CrossRef]

32. Dudenbostel, T.; Acelajado, M.C.; Pisoni, R.; Li, P.; Oparil, S.; Calhoun, D.A. Refractory Hypertension: Evidence of Heightened Sympathetic Activity as a Cause of Antihypertensive Treatment Failure. *Hypertension* **2015**, *66*, 126–133. [CrossRef]

33. Dudenbostel, T.; Siddiqui, M.; Gharpure, N.; Calhoun, D.A. Refractory versus resistant hypertension: Novel distinctive phenotypes. *J. Nat. Sci.* **2017**, *3*, e430.

34. Inui, K.; Nosaka, S. Target site of inhibition mediated by midbrain periaqueductal gray matter of baroreflex vagal bradycardia. *J. Neurophysiol.* **1993**, *70*, 2205–2214. [CrossRef]

35. Lovick, T. Inhibitory modulation of the cardiovascular defence response by the ventrolateral periaqueductal grey matter in rats. *Exp. Brain Res.* **1992**, *89*, 133–139. [CrossRef]

36. Hyam, J.A.; Kringelbach, M.L.; Silburn, P.A.; Aziz, T.Z.; Green, A.L. The autonomic effects of deep brain stimulation—A therapeutic opportunity. *Nat. Rev. Neurol.* **2012**, *8*, 391–400. [CrossRef]

37. Juraschek, S.P.; Daya, N.; Appel, L.J.; Miller, E.R.; Windham, B.G.; Pompeii, L.; Griswold, M.E.; Kucharska-Newton, A.; Selvin, E. Orthostatic Hypotension in Middle-Age and Risk of Falls. *Am. J. Hypertens.* **2017**, *30*, 188–195. [CrossRef]

38. Basnayake, S.D.; Hyam, J.A.; Pereira, E.A.; Schweder, P.M.; Brittain, J.S.; Aziz, T.Z.; Green, A.L. Identifying cardiovascular neurocircuitry involved in the exercise pressor reflex in humans using functional neurosurgery. *J. Appl. Physiol.* **2011**, *110*, 881–891. [CrossRef]

39. Ricci, F.; De Caterina, R.; Fedorowski, A. Orthostatic Hypotension: Epidemiology, Prognosis, and Treatment. *J. Am. Coll. Cardiol.* **2015**, *66*, 848–860. [CrossRef]

40. Krygowska-Wajs, A.; Furgala, A.; Gorecka-Mazur, A.; Pietraszko, W.; Thor, P.; Potasz-Kulikowsja, K.; Moskala, M. The effect of subthalamic deep brain stimulation on gastric motility in Parkinson's disease. *Parkinsonism Relat. Disord.* **2016**, *26*, 35–40. [CrossRef]

41. Green, A.L.; Wang, S.; Bittar, R.G.; Owen, S.L.; Paterson, D.J.; Stein, J.F.; Bain, P.G.; Shlugman, D.; Aziz, T.Z. Deep brain stimulation: A new treatment for hypertension? *J. Clin. Neurosci.* **2007**, *14*, 592–595. [CrossRef]

42. Holmberg, B.; Corneliusson, O.; Elam, M. Bilateral stimulation of nucleus subthalamicus in advanced Parkinson's disease: No effects on, and of, autonomic dysfunction. *Mov. Disord.* **2005**, *20*, 976–981. [CrossRef]

43. Green, A.L.; Hyam, J.A.; Williams, C.; Wang, S.; Shlugman, D.; Stein, J.F.; Paterson, D.J.; Aziz, T.Z. Intra-operative deep brain stimulation of the periaqueductal grey matter modulates blood pressure and heart rate variability in humans. *Neuromodulation* **2010**, *13*, 174–181. [CrossRef]

44. Abrahams, V.C.; Hilton, S.M.; Zbrozyna, A. Active muscle vasodilatation produced by stimulation of the brain stem: Its significance in the defence reaction. *J. Physiol.* **1960**, *154*, 491–513. [CrossRef] [PubMed]

45. Carrive, P.; Bandler, R. Control of extra cranial and hindlimb blood flow by the midbrain periaqueductal grey of the cat. *Exp. Brain Res.* **1991**, *84*, 599–606. [CrossRef] [PubMed]

46. Lovick, T. Deep brain stimulation and autonomic control. *Exp. Physiol.* **2014**, *99*, 320–325. [CrossRef] [PubMed]

47. Green, A.L.; Wang, S.; Owen, S.L.; Xie, K.; Liu, X.; Paterson, D.J.; Stein, J.F.; Bain, P.G.; Aziz, T.Z. Deep brain stimulation can regulate arterial blood pressure in awake humans. *Neuroreport* **2005**, *16*, 1741–1745. [CrossRef] [PubMed]

48. O'Callaghan, E.L.; Hart, E.C.; Sims-Williams, H.; Javed, S.; Burchell, A.E.; Papouchado, M.; Tank, J.; Heusser, K.; Jordan, J.; Menne, J.; et al. Chronic Deep Brain Stimulation Decreases Blood Pressure and Sympathetic Nerve Activity in a Drug- and Device-Resistant Hypertensive Patient. *Hypertension* **2017**, *69*, 522–528. [CrossRef] [PubMed]

49. Patel, N.K.; Javed, S.; Khan, S.; Papouchado, M.; Malizia, A.L.; Pickering, A.E.; Paton, J.F. Deep brain stimulation relieves refractory hypertension. *Neurology* **2011**, *76*, 405–407. [CrossRef] [PubMed]

50. Pereira, E.A.; Wang, S.; Paterson, D.J.; Stein, J.F.; Aziz, T.Z.; Green, A.L. Sustained reduction of hypertension by deep brain stimulation. *J. Clin. Neurosci.* **2010**, *17*, 124–127. [CrossRef] [PubMed]

51. Green, A.L.; Wang, S.; Owen, S.L.; Paterson, D.J.; Stein, J.F.; Aziz, T.Z. Controlling the heart via the brain: A potential new therapy for orthostatic hypotension. *Neurosurgery* **2006**, *58*, 1176–1183, discussion 1176–1183. [CrossRef] [PubMed]

52. Ludwig, J.; Remien, P.; Guballa, C.; Binder, A.; Binder, S.; Schattschneider, J.; Herzog, J.; Volkmann, J.; Deuschl, G.; Wasner, G. Effects of subthalamic nucleus stimulation and levodopa on the autonomic nervous system in Parkinson's disease. *J. Neurol. Neurosurg. Psychiatry* **2007**, *78*, 742–745. [CrossRef]

53. Stemper, B.; Beric, A.; Welsch, G.; Haendl, T.; Sterio, D.; Hilz, M.J. Deep brain stimulation improves orthostatic regulation of patients with Parkinson disease. *Neurology* **2006**, *67*, 1781–1785. [CrossRef]

54. Sverrisdóttir, Y.B.; Green, A.L.; Aziz, T.Z.; Bahuri, N.F.; Hyam, J.; Basnayake, S.D.; Paterson, D.J. Differentiated baroreflex modulation of sympathetic nerve activity during deep brain stimulation in humans. *Hypertension* **2014**, *63*, 1000–1010. [CrossRef] [PubMed]

55. Thornton, J.M.; Aziz, T.; Schlugman, D.; Paterson, D.J. Electrical stimulation of the midbrain increases heart rate and arterial blood pressure in awake humans. *J. Physiol.* **2002**, *539 Pt 2*, 615–621. [CrossRef]

56. Trachani, E.; Constantoyannis, C.; Sakellaropoulos, G.C.; Stavrinou, M.L.; Nikiforidis, G.; Chroni, E. Heart rate variability in Parkinson's disease unaffected by deep brain stimulation. *Acta Neurol. Scand.* **2012**, *126*, 56–61. [CrossRef] [PubMed]

57. Furgała, A.; Górecka-Mazur, A.; Fiszer, U.; Pietraszko, W.; Thor, P.; Moskala, M.; Potasz, K.; Bukowczan, M.; Polak, J.; Krygowska-Wajs, A. Evaluation of heart rate and blood pressure variability in Parkinson's disease patients after bilateral subthalamic deep brain stimulation. *Przegl Lek.* **2015**, *72*, 246–252.

58. Sumi, K.; Katayama, Y.; Otaka, T.; Obuchi, T.; Kano, K.; Kobayashi, K.; Oshima, H.; Fukaya, C.; Yamamoto, T.; Ogawa, Y. Effect of subthalamic nucleus deep brain stimulation on the autonomic nervous system in Parkinson's disease patients assessed by spectral analyses of R-R interval variability and blood pressure variability. *Stereotact. Funct. Neurosurg.* **2012**, *90*, 248–254. [CrossRef]

59. Ambrosino, N.; Simonds, A. The clinical management in extremely severe COPD. *Respir. Med.* **2007**, *101*, 1613–1624. [CrossRef] [PubMed]

60. Zielinski, J.; Bednarek, M.; Górecka, D.; Viegi, G.; Hurd, S.S.; Fukuchi, Y.; Lai, C.K.; Ran, P.X.; Ko, F.W.; Liu, S.M. Increasing COPD awareness. *Eur. Respir. J.* **2006**, *27*, 833–852. [CrossRef]

61. Spencer, W.G. The effect produced upon respiration by faraday excitation of the cerebrum in the monkey, dog, cat and rabbit. *Philos. Trans. R. Soc. Lond. B* **1894**, *185*, 609–657.

62. Pool, J.L.; Ransohoff, J. Autonomic effects on stimulating rostral portion of cingulate gyri in man. *J. Neurophysiol.* **1949**, *12*, 385–392. [CrossRef]

63. Hyam, J.A.; Brittain, J.S.; Paterson, D.J.; Davies, R.J.; Aziz, T.Z.; Green, A.L. Controlling the lungs via the brain: A novel neurosurgical method to improve lung function in humans. *Neurosurgery* **2012**, *70*, 469–477, discussion 477–468. [CrossRef] [PubMed]

64. Bingol, H.; Cingoz, F.; Balkan, A.; Kilic, S.; Bolcal, C.; Demirkilic, U.; Tatar, H. The effect of oral prednisolone with chronic obstructive pulmonary disease undergoing coronary artery bypass surgery. *J. Card. Surg.* **2005**, *20*, 252–256. [CrossRef] [PubMed]

65. Li, H.; He, G.; Chu, H.; Zhao, L.; Yu, H. A step-wise application of methylprednisolone versus dexamethasone in the treatment of acute exacerbations of COPD. *Respirology* **2003**, *8*, 199–204. [CrossRef] [PubMed]

66. Ställberg, B.; Selroos, O.; Vogelmeier, C.; Andersson, E.; Ekström, T.; Larsson, K. Budesonide/formoterol as effective as prednisolone plus formoterol in acute exacerbations of COPD. A double-blind, randomised, non-inferiority, parallel-group, multicentre study. *Respir. Res.* **2009**, *10*, 11. [CrossRef] [PubMed]

67. Aron, A.R.; Poldrack, R.A. Cortical and subcortical contributions to Stop signal response inhibition: Role of the subthalamic nucleus. *J. Neurosci.* **2006**, *26*, 2424–2433. [CrossRef] [PubMed]
68. Ray, N.J.; Jenkinson, N.; Kringelbach, M.L.; Hansen, P.C.; Pereira, E.A.; Brittain, J.S.; Holland, P.; Holliday, I.E.; Owen, S.; Stein, J.; et al. Abnormal thalamocortical dynamics may be altered by deep brain stimulation: Using magnetoencephalography to study phantom limb pain. *J. Clin. Neurosci.* **2009**, *16*, 32–36. [CrossRef] [PubMed]
69. Bittencourt, A.S.; Carobrez, A.P.; Zamprogno, L.P.; Tufik, S.; Schenberg, L.C. Organization of single components of defensive behaviors within distinct columns of periaqueductal gray matter of the rat: Role of N-methyl-D-aspartic acid glutamate receptors. *Neuroscience* **2004**, *125*, 71–89. [CrossRef] [PubMed]
70. Carrive, P.; Bandler, R. Viscerotopic organization of neurons subserving hypotensive reactions within the midbrain periaqueductal grey: A correlative functional and anatomical study. *Brain Res.* **1991**, *541*, 206–215. [CrossRef]
71. McGaraughty, S.; Farr, D.A.; Heinricher, M.M. Lesions of the periaqueductal gray disrupt input to the rostral ventromedial medulla following microinjections of morphine into the medial or basolateral nuclei of the amygdala. *Brain Res.* **2004**, *1009*, 223–227. [CrossRef]
72. Gautier, H.; Bertrand, F. Respiratory effects of pneumotaxic center lesions and subsequent vagotomy in chronic cats. *Respir. Physiol.* **1975**, *23*, 71–85. [CrossRef]
73. Mraovitch, S.; Kumada, M.; Reis, D.J. Role of the nucleus parabrachialis in cardiovascular regulation in cat. *Brain Res.* **1982**, *232*, 57–75. [CrossRef]
74. Motekaitis, A.M.; Solomon, I.C.; Kaufman, M.P. Stimulation of parabrachial nuclei dilates airways in cats. *J. Appl. Physiol.* **1994**, *76*, 1712–1718. [CrossRef]
75. Motekaitis, A.M.; Solomon, I.C.; Kaufman, M.P. Blockade of glutamate receptors in CVLM and NTS attenuates airway dilation evoked from parabrachial region. *J. Appl. Physiol.* **1996**, *81*, 400–407. [CrossRef]
76. American Thoracic Society. Thoracic Society. Dyspnea. Mechanisms, assessment, and management: A consensus statement. American Thoracic Society. *Am. J. Respir. Crit. Care Med.* **1999**, *159*, 321–340. [CrossRef]
77. Sakakibara, R.; Shinotoh, H.; Uchiyama, T.; Sakuma, M.; Kashiwado, M.; Yoshiyama, M.; Hattori, T. Questionnaire-based assessment of pelvic organ dysfunction in Parkinson's disease. *Auton. Neurosci.* **2001**, *92*, 76–85. [CrossRef]
78. Sakakibara, R.; Tateno, F.; Nagao, T.; Yamamoto, T.; Uchiyama, T.; Yamanishi, M.; Yano, M.; Kishi, M.; Tsuyusaki, Y.; Aiba, Y. Bladder function of patients with Parkinson's disease. *Int. J. Urol.* **2014**, *21*, 638–646. [CrossRef]
79. Sakakibara, R.; Panicker, J.; Finazzi-Agro, E.; Iacovelli, V.; Bruschini, H.; Parkinson's Disease Subcomittee TeNPCiTICS. A guideline for the management of bladder dysfunction in Parkinson's disease and other gait disorders. *Neurourol. Urodyn.* **2016**, *35*, 551–563. [CrossRef] [PubMed]
80. Winge, K.; Skau, A.M.; Stimpel, H.; Nielsen, K.K.; Werdelin, L. Prevalence of bladder dysfunction in Parkinsons disease. *Neurourol. Urodyn.* **2006**, *25*, 116–122. [CrossRef]
81. Roy, H.A.; Aziz, T.Z. Deep brain stimulation and multiple sclerosis: Therapeutic applications. *Mult. Scler. Relat. Disord.* **2014**, *3*, 431–439. [CrossRef]
82. Winge, K.; Nielsen, K.K. Bladder dysfunction in advanced Parkinson's disease. *Neurourol. Urodyn.* **2012**, *31*, 1279–1283. [CrossRef]
83. Kessler, T.M.; La Framboise, D.; Trelle, S.; Fowler, C.J.; Kiss, G.; Pannek, J.; Schurch, B.; Sievert, K.D.; Engeler, D.S. Sacral neuromodulation for neurogenic lower urinary tract dysfunction: Systematic review and meta-analysis. *Eur. Urol.* **2010**, *58*, 865–874. [CrossRef]
84. Gómez-Pinilla, P.J.; Pozo, M.J.; Camello, P.J. Aging impairs neurogenic contraction in guinea pig urinary bladder: Role of oxidative stress and melatonin. *Am. J. Physiol. Regul. Integr. Comp. Physiol.* **2007**, *293*, R793–R803. [CrossRef]
85. Ju, Y.H.; Liao, L.M. Electrical stimulation of dog pudendal nerve regulates the excitatory pudendal-to-bladder reflex. *Neural Regen. Res.* **2016**, *11*, 676–681.
86. Blok, B.F.; de Weerd, H.; Holstege, G. The pontine micturition center projects to sacral cord GABA immunoreactive neurons in the cat. *Neurosci. Lett.* **1997**, *233*, 109–112. [CrossRef]
87. Blok, B.F.; Holstege, G. The pontine micturition center in rat receives direct lumbosacral input. An ultrastructural study. *Neurosci. Lett.* **2000**, *282*, 29–32. [CrossRef]

88. Sasa, M.; Yoshimura, N. Locus coeruleus noradrenergic neurons as a micturition center. *Microsc. Res. Tech.* **1994**, *29*, 226–230. [CrossRef]

89. Dalmose, A.L.; Bjarkam, C.R.; Sørensen, J.C.; Djurhuus, J.C.; Jørgensen, T.M. Effects of high frequency deep brain stimulation on urine storage and voiding function in conscious minipigs. *Neurourol. Urodyn.* **2004**, *23*, 265–272. [CrossRef]

90. Finazzi-Agrò, E.; Peppe, A.; D'Amico, A.; Petta, F.; Mazzone, P.; Stanzione, P.; Micali, F.; Caltagirone, C. Effects of subthalamic nucleus stimulation on urodynamic findings in patients with Parkinson's disease. *J. Urol.* **2003**, *169*, 1388–1391. [CrossRef]

91. Herzog, J.; Weiss, P.H.; Assmus, A.; Wefer, B.; Seif, C.; Braum, P.M.; Herzog, H.; Volkmann, J.; Deuschl, G.; Fink, G.R. Subthalamic stimulation modulates cortical control of urinary bladder in Parkinson's disease. *Brain* **2006**, *129 Pt 12*, 3366–3375. [CrossRef]

92. Seif, C.; Herzog, J.; van der Horst, C.; Schrader, B.; Volkmann, J.; Deuschl, G.; Juenemann, K.P.; Braun, P.M. Effect of subthalamic deep brain stimulation on the function of the urinary bladder. *Ann. Neurol.* **2004**, *55*, 118–120. [CrossRef]

93. Green, A.L.; Stone, E.; Sitsapesan, H.; Turney, B.W.; Coote, J.H.; Aziz, T.Z.; Hyam, J.A.; Lovick, T.A. Switching off micturition using deep brain stimulation at midbrain sites. *Ann. Neurol.* **2012**, *72*, 144–147. [CrossRef]

94. Kessler, T.M.; Burkhard, F.C.; Z'Brun, S.; Stibal, A.; Studer, U.E.; Hess, C.W.; Kaelin-Lang, A. Effect of thalamic deep brain stimulation on lower urinary tract function. *Eur. Urol.* **2008**, *53*, 607–612. [CrossRef]

95. Skultety, F.M. Relation to periaqueductal gray matter to stomach and bladder motility. *Neurology* **1959**, *9*, 190–198. [CrossRef]

96. Blok, B.F.; De Weerd, H.; Holstege, G. Ultrastructural evidence for a paucity of projections from the lumbosacral cord to the pontine micturition center or M-region in the cat: A new concept for the organization of the micturition reflex with the periaqueductal gray as central relay. *J. Comp. Neurol.* **1995**, *359*, 300–309. [CrossRef]

97. Blok, B.F.; Holstege, G. Direct projections from the periaqueductal gray to the pontine micturition center (M-region). An anterograde and retrograde tracing study in the cat. *Neurosci. Lett.* **1994**, *166*, 93–96. [CrossRef]

98. Tai, C.; Wang, J.; Jin, T.; Wang, P.; Kim, S.G.; Roppolo, J.R.; de Groat, W.C. Brain switch for reflex micturition control detected by FMRI in rats. *J. Neurophysiol.* **2009**, *102*, 2719–2730. [CrossRef]

99. Stone, E.; Coote, J.H.; Lovick, T.A. Effect of electrical vs. chemical deep brain stimulation at midbrain sites on micturition in anaesthetized rats. *Acta Physiol.* **2015**, *214*, 135–145. [CrossRef]

100. Stone, E.; Coote, J.H.; Allard, J.; Lovick, T.A. GABAergic control of micturition within the periaqueductal grey matter of the male rat. *J. Physiol.* **2011**, *589 Pt 8*, 2065–2078. [CrossRef]

101. An, X.; Bandler, R.; Ongür, D.; Price, J.L. Prefrontal cortical projections to longitudinal columns in the midbrain periaqueductal gray in macaque monkeys. *J. Comp. Neurol.* **1998**, *401*, 455–479. [CrossRef]

102. Fowler, C.J.; Griffiths, D.; de Groat, W.C. The neural control of micturition. *Nat. Rev. Neurosci.* **2008**, *9*, 453–466. [CrossRef]

103. Holstege, G. The emotional motor system and micturition control. *Neurourol. Urodyn.* **2010**, *29*, 42–48. [CrossRef]

104. Aviles-Olmos, I.; Foltynie, T.; Panicker, J.; Cowie, D.; Limousin, P.; Hariz, M.; Fowler, C.J.; Zrinzo, L. Urinary incontinence following deep brain stimulation of the pedunculopontine nucleus. *Acta Neurochir.* **2011**, *153*, 2357–2360. [CrossRef] [PubMed]

105. Ferraye, M.U.; Debû, B.; Fraix, V.; Goetz, L.; Ardouin, C.; Yelnik, J.; Henry-Lagrange, C.; Seigneuret, E.; Piallat, B.; Krack, P. Effects of pedunculopontine nucleus area stimulation on gait disorders in Parkinson's disease. *Brain* **2010**, *133 Pt 1*, 205–214. [CrossRef]

106. Nandi, D.; Jenkinson, N.; Stein, J.; Aziz, T. The pedunculopontine nucleus in Parkinson's disease: Primate studies. *Br. J. Neurosurg.* **2008**, *22* (Suppl. 1), S4–S8. [CrossRef] [PubMed]

107. Plaha, P.; Gill, S.S. Bilateral deep brain stimulation of the pedunculopontine nucleus for Parkinson's disease. *Neuroreport* **2005**, *16*, 1883–1887. [CrossRef] [PubMed]

108. Stefani, A.; Lozano, A.M.; Peppe, A.; Stanzione, P.; Galati, S.; Tropepi, D.; Pierantozzi, M.; Brusa, L.; Scarnati, E.; Mazzone, P. Bilateral deep brain stimulation of the pedunculopontine and subthalamic nuclei in severe Parkinson's disease. *Brain* **2007**, *130 Pt 6*, 1596–1607. [CrossRef]

109. Jensen, K.N.; Deding, D.; Sørensen, J.C.; Bjarkam, C.R. Long-term implantation of deep brain stimulation electrodes in the pontine micturition centre of the Göttingen minipig. *Acta Neurochir.* **2009**, *151*, 785–794, discussion 794. [CrossRef]

110. Roy, H.A.; Pond, D.; Roy, C.; Forrow, B.; Foltynie, L.; Zrinzo, L.; Akram, H.; Aziz, T.Z.; FitzGerald, J.J.; Green, A.L. Effects of pedunculopontine nucleus stimulation on human bladder function. *Neurourol. Urodyn.* **2018**, *37*, 726–734. [CrossRef]

111. Chen, S.C.; Chu, P.Y.; Hsieh, T.H.; Li, Y.T.; Peng, C.W. Feasibility of deep brain stimulation for controlling the lower urinary tract functions: An animal study. *Clin. Neurophysiol.* **2017**, *128*, 2438–2449. [CrossRef]

112. Witjas, T.; Kaphan, E.; Régis, J.; Jouve, E.; CheriF, A.A.; Peragut, J.C.; Azulay, J.P. Effects of chronic subthalamic stimulation on nonmotor fluctuations in Parkinson's disease. *Mov. Disord.* **2007**, *22*, 1729–1734. [CrossRef]

113. Halim, A.; Baumgartner, L.; Binder, D.K. Effect of deep brain stimulation on autonomic dysfunction in patients with Parkinson's disease. *J. Clin. Neurosci.* **2011**, *18*, 804–806. [CrossRef]

114. Trachani, E.; Constantoyannis, C.; Sirrou, V.; Kefalopoulou, Z.; Markaki, E.; Chroni, E. Effects of subthalamic nucleus deep brain stimulation on sweating function in Parkinson's disease. *Clin. Neurol. Neurosurg.* **2010**, *112*, 213–217. [CrossRef]

115. Diamond, A.; Kenney, C.; Almaguer, M.; Jankovic, J. Hyperhidrosis due to deep brain stimulation in a patient with essential tremor. Case report. *J. Neurosurg.* **2007**, *107*, 1036–1038. [CrossRef]

116. Ciucci, M.R.; Barkmeier-Kraemer, J.M.; Sherman, S.J. Subthalamic nucleus deep brain stimulation improves deglutition in Parkinson's disease. *Mov. Disord.* **2008**, *23*, 676–683. [CrossRef]

117. Derrey, S.; Chastan, N.; Maltete, D.; Verin, E.; Dechelotte, R.; Lefaucher, R.; Proust, F.; Freger, P.; Leroi, A.M.; Weber, J.; et al. Impact of deep brain stimulation on pharyngo-esophageal motility: A randomized cross-over study. *Neurogastroenterol. Motil.* **2015**, *27*, 1214–1222. [CrossRef]

118. Silbergleit, A.K.; LeWitt, P.; Junn, F.; Schiltz, L.R.; Collins, D.; Beardsley, T.; Hubert, M.; Trosch, R.; Schwalb, J.M. Comparison of dysphagia before and after deep brain stimulation in Parkinson's disease. *Mov. Disord.* **2012**, *27*, 1763–1768. [CrossRef]

119. Arai, E.; Arai, M.; Uchiyama, T.; Higuchi, Y.; Aoyagi, K.; Yamanaka, Y.; Yamamoto, T.; Nagano, O.; Shiina, A.; Maruoka, D.; et al. Subthalamic deep brain stimulation can improve gastric emptying in Parkinson's disease. *Brain* **2012**, *135 Pt 5*, 1478–1485. [CrossRef]

120. Becerra, J.E.; Zorro, O.; Ruiz-Gaviria, R.; Cataneda-Cardona, C.; Otalora-Esteban, M.; Henao, S.; Navarrete, S.; Acevedo, J.C.; Rosselli, D. Economic Analysis of Deep Brain Stimulation in Parkinson Disease: Systematic Review of the Literature. *World Neurosurg.* **2016**, *93*, 44–49. [CrossRef]

121. Garrard, E. Palliative care and the ethics of resource allocation. *Int. J. Palliat. Nurs.* **1996**, *2*, 91–94. [CrossRef]

122. Jaakkimainen, L.; Goodwin, P.J.; Pater, J.; Warde, P.; Murray, N.; Rapp, E. Counting the costs of chemotherapy in a National Cancer Institute of Canada randomized trial in nonsmall-cell lung cancer. *J. Clin. Oncol.* **1990**, *8*, 1301–1309. [CrossRef]

123. Kaplan, N.M. The promises and perils of treating the elderly hypertensive. *Am. J. Med. Sci.* **1993**, *305*, 183–197. [CrossRef]

124. Pereira, E.A.; Paranathala, M.; Hyam, J.A.; Green, A.L.; Aziz, T.Z. Anterior cingulotomy improves malignant mesothelioma pain and dyspnoea. *Br. J. Neurosurg.* **2014**, *28*, 471–474. [CrossRef]

Review

Neuromodulation for Pelvic and Urogenital Pain

Holly Roy [1], Ifeoma Offiah [2] and Anu Dua [2,*]

[1]   Neurosurgery Department, University Hospitals Plymouth, Plymouth PL6 8DH, UK; holly.roy@nhs.net
[2]   Department of Obstetrics and Gynaecology, University Hospitals Plymouth, Plymouth PL6 8DH, UK;
      ifyoffiah@nhs.net
*   Correspondence: anudua@nhs.net; Tel.: +44-7387-850745

Received: 14 August 2018; Accepted: 28 September 2018; Published: 29 September 2018

Abstract: Chronic pain affecting the pelvic and urogenital area is a major clinical problem with heterogeneous etiology, affecting both male and female patients and severely compromising quality of life. In cases where pharmacotherapy is ineffective, neuromodulation is proving to be a potential avenue to enhance analgesic outcomes. However, clinicians who frequently see patients with pelvic pain are not traditionally trained in a range of neuromodulation techniques. The aim of this overview is to describe major types of pelvic and urogenital pain syndromes and the neuromodulation approaches that have been trialed, including peripheral nerve stimulation, dorsal root ganglion stimulation, spinal cord stimulation, and brain stimulation techniques. Our conclusion is that neuromodulation, particularly of the peripheral nerves, may provide benefits for patients with pelvic pain. However, larger prospective randomized studies with carefully selected patient groups are required to establish efficacy and determine which patients are likely to achieve the best outcomes.

Keywords: pelvic pain; bladder-pain syndrome; neuromodulation; posterior tibial-nerve stimulation; sacral-nerve stimulation; dorsal-root-ganglion stimulation

---

1. Introduction

1.1. Pelvic Anatomy and Pelvic and Urogenital Pain Syndromes

Pain derived from the pelvic and urogenital region is an important clinical problem affecting both men and women. Patients may present to a range of different clinical specialties including gynecology, urology, and general surgery. Pelvic and urogenital pain syndromes include chronic pelvic pain/chronic prostatitis (CPP/CP), bladder-pain syndrome (BPS), groin/inguinal pain, and genital pain. We begin by revising relevant pelvic and urogenital anatomy before introducing the pathophysiology of individual pain syndromes.

The true pelvis is the anatomical area between the floor of the pelvic cavity (composed of the pelvic and urogenital diaphragms) and the pelvic brim. Organs occupying the pelvis include the urinary bladder and the uterus in their empty states, the rectum, vagina, and distal parts of the male reproductive system. Both visceral and somatic nerves innervate structures within the pelvis; the innervation is complex, and in this section we will briefly summarize the main pathways for afferent (sensory) information transmission [1] (see Figure 1). Sensory afferent information from the colon, bladder, and urethra is transmitted via the splanchnic, pudendal, and pelvic nerves, whose cell bodies exist in dorsal root ganglia (DRG) at the level of the lumbosacral and thoracolumbar cord [2]. Somatic sensation to the clitoris/penis, perineal skin, and distal aspect of the anal canal are provided by branches of the pudendal nerve [3–5]. Furthermore, the ilioinguinal, genitofemoral, and iliohypogastric nerves provide overlapping innervation of the skin in the groin/pubic region [5].

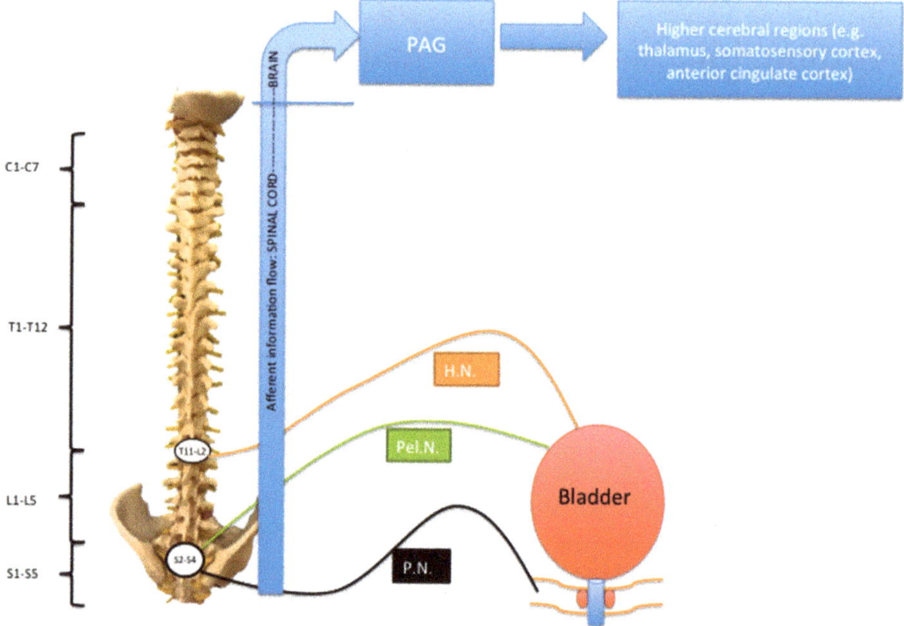

Figure 1. P.N.: pudendal nerve; Pel.N.: pelvic nerve; H.N. hypogastric nerve; PAG: periaqueductal grey area. Schematic to summarize afferent innervation of the lower urinary tract. The sensory fibers traveling in the pelvic and pudendal nerves have their cell bodies in dorsal root ganglia (DRGs) at the S2–S4 level. Parasympathetic fibers travel in the pelvic nerve and sympathetic fibers travel in the hypogastric nerve. Modified from Reference [1].

Pain from the bladder is transmitted by visceral afferent nerves travelling with the sympathetic nerves via the hypogastric nerve, and also, in the case of the lower segment of the bladder, sacral parasympathetics in the pelvic nerve. Pain from the upper pelvic viscera accompanies sympathetics. Pain from the lower viscera, such as the cervix and upper vagina, travels with the parasympathetic fibers.

Visceral afferents synapse onto second-order neurons in the dorsal horn of the spinal cord. There may be convergence of either visceral or somatic inputs onto the second-order neuron, which may potentially give rise to referred pain or cross-sensitization [6]. From the second-order neuron, information then passes either along the spinothalamic tract or the dorsal column medial lemniscus pathways to supraspinal regions responsible for processing the affective and sensory components of pain, including the periaqueductal grey area, thalamus, somatosensory cortex, and anterior cingulate cortex [7,8].

Chronic pelvic pain localized to the bladder, genitals, groin, or anorectum may be a direct result of nerve injury, inflammation, or entrapment, or may have a secondary neural component that contributes to pain amplification or maintenance. Important nerves to consider include the pudendal, thoracolumbar, ilioinguinal, iliohypogastric, genitofemoral, or obturator nerve [9]. Imaging techniques, including magnetic resonance neurography, are becoming increasingly valuable in diagnosing these conditions [10].

Afferent nociceptive plasticity and long-term plasticity in the dorsal horn of the spinal cord and supraspinal regions are important events underlying the development of chronic pain [2,8], in which the experience of pain persists after initial tissue damage has healed, and the pain has additional components such as spontaneous pain, hyperalgesia, and allodynia [7]. Many factors are involved in the development of chronic pain, including peripheral afferent sensitization, and long-term synaptic and molecular changes within the dorsal horn and brain [2,7,8,11–13]. Interaction with the immune system, including the microglial response, is also considered to be important in the transition to a chronic pain state [7]. Further characterization of the molecular, cellular, and network changes involved in the development of the chronic pain state are key in determining future approaches to treatment and the role of neuromodulation [7,14,15].

1.2. Major Pelvic and Urogenital Pain Syndromes (Should Include Epidemiology, Pathophysiology)

Chronic pelvic pain/chronic prostatitis (CPP/CP) is a complex pain syndrome of heterogeneous etiology. There are many inconsistencies with the way in which CPP/CP has been reported and defined, and for that reason, it is not easy to quote an accurate figure for its true incidence and prevalence [16]. However, presence of pain for at least six months is generally considered to be necessary for a diagnosis to be considered, and as a rough estimate, CPP affects 38/1000 women annually in the United Kingdom (UK), or has a prevalence of 1 million women in the UK [16], and 9.2 million women in the United States [17,18]. CP affects men, and involves symptoms of pelvic pain and/or bothersome symptoms when urinating. The U.K. prevalence of chronic prostatitis has been estimated as 8.2% [16]. A recent article supported by the International Continence Society (ICS) described nine domains to be used for the description of chronic pelvic pain, including four pelvic organ domains ((1) lower urinary tract; (2) female genital; (3) male genital; and (4) gastrointestinal); two domains representing pain perceived in the pelvis but not necessarily arising from the pelvic organs ((5) musculoskeletal; (6) neurological); and three domains that may influence pain perception ((7) psychological; (8) sexual; (9) comorbidities) [9]. In this article, we concentrate primarily on the first three pelvic-organ domains (lower urinary tract, female genital, and male genital), and the role of neurological factors in the development of pain in these domains. However, the contribution of musculoskeletal, psychological, sexual and other disease factors should not be ignored by the physician caring for patients with pelvic pain.

Chronic pain experienced in the lower-urinary-tract domain refers primarily to bladder pain syndrome (BPS)/interstitial cystitis (IC). In this article, we will refer to it as BPS. This has been defined previously by the ICS as pain or discomfort related to the urinary bladder, which is associated with other urinary symptoms, such as frequency and urgency, with the exclusion of any other diseases of the lower urinary tract [19]. Prevalence reports vary depending on the country of origin and diagnostic criteria, but are in the range of 3–4 per 100,000 in Japan to 450 per 100,000 in the Finnish population [20,21]. The precise trigger resulting in the development of BPS is still unknown. However, it is possible that bladder injury by irritant chemicals, radiation, blunt trauma, childbirth, or subclinical infection may trigger the release of inflammatory mediators and consequent disruption of the protective mucosal barrier [22,23]. Resident and recruited immune cells as well as toxic urinary solutes permeate the barrier and lead to depolarization of sensory afferents, causing bladder pain.

Chronic pelvic pain in the male or female genital domains may be localized to the vagina, vulva, or perineum, or may involve intra-abdominal organs, including ovaries, uterus, and fallopian tubes (females), and can involve the prostate, scrotum, epididymis, testicles, or penis (males). Endometriosis, adenomyosis, adhesions, pelvic inflammatory disease, pelvic masses, peripheral pelvic neuropathies, and Tarlov cysts are potential causes [5,24,25]. Pelvic pain arising specifically from entrapment or neuropathy of the pudendal nerve is known as pudendal neuralgia (PN), which results in chronic perineal pain. Pain can extend from the perianal region to the vicinity of the scrotum/clitoris anteriorly [3]. The diagnostic criteria for pudendal neuralgia as described by the Nantes criteria include: (i) pain in the distribution of the pudendal nerve; (ii) pain experienced most significantly when sitting; (iii) pain that does not wake the patient at night; (iv) pain that is not associated with an objective sensory impairment and; (v) pain relieved by diagnostic pudendal nerve block [26]. Genital pain can also develop following lower abdominal or pelvic surgery, as in the case of scrotal pain following vasectomy [27], or, rarely, clitoral pain following midurethral sling placement [28], which is likely to be related to intraoperative nerve trauma. Pelvic pain can localize to the groin area, which may develop as a complication of inguinal hernia repair [29,30] where direct nerve damage, neuroma, postsurgical fibrosis, or compression can occur [7], resulting in pain radiating into the groin, thigh, or genitals. It is thought that up to 12% of patients may experience pain that impairs daily activity after hernia repair [31].

Chronic pelvic pain in the gastrointestinal domain also includes pain in the anorectal area. This can be a result of structural problems such as abscesses, anal fissures, cryptitis, and hemorrhoids [9,32], or conditions of other etiologies, such as chronic proctalgia, which may be related to pelvic floor muscle hypertonicity. Associated symptoms include diarrhea, constipation, abdominal cramps, and rectal pain. There may be associated bladder and urethral symptoms due to cross-sensitization between the bladder and colon [2]. Recognizing the possible link between colonic inflammation and bladder pain, and vice versa, is important when approaching the problem of pelvic pain.

1.3. Pharmacological and Non-Neuromodulatory Surgical Interventions for Pelvic and Urogenital Pain Syndromes

Initial approaches for the management of pelvic and urogenital pain syndromes are conservative and include physical therapy where indicated, simple analgesia such as paracetamol and nonsteroidal anti-inflammatories, neuropathic analgesics (particularly in patients whose pain appears to be neuropathic in nature, with a "burning" or "stabbing" quality and/or in the distribution of a known peripheral nerve) including tricyclic antidepressants, pregabalin, gabapentin, selective serotonin reuptake inhibitors, and N-methyl-D-aspartate receptor antagonists, or opioid analgesia [33,34]. Other pharmacological agents utilized for the treatment of chronic pelvic pain in women include hormonal agents (goserelin, medroxyprogesterone acetate), venoconstrictors such as ergotamine, and venomimetics such as daflon [34], while in men with chronic prostatitis, antibiotics, 5-alpha-reductase inhibitors (if benign prostatic hyperplasia is present), and phytotherapy may be used [33,35]. Psychological therapy may also be offered if indicated. Subsequent approaches in patients who are resistant to initial management and trial of pharmacotherapy tends to vary according to the pain syndrome. For bladder pain syndrome, subsequent therapy may include replacing the urothelial barrier using intravesical installation of glycosaminoglycans (such as pentosan polysulfate sodium or hyaluronic acid) [36], or use of botulinum toxin. Nerve blocks may be used if the pain is thought to be neurogenic in origin [3,5].

Despite a range of conservative and pharmacological options for the management of chronic pelvic and urogenital pain, (see Reference [34] for a review of specific management options for individual pelvic-pain conditions) there remains a group of patients who are resistant to pharmacological interventions. It is usually this patient group that is considered for neuromodulation, particularly if they have shown short-term responsiveness to nerve blocks. The use of neuromodulation for various chronic pelvic pain syndromes is still in its experimental phase and a matter of considerable debate [16]. Neuromodulation ranges from peripheral nerve stimulation, usually using percutaneous electrodes to target a peripheral nerve, to dorsal root ganglion stimulation, spinal cord stimulation, and brain stimulation (see Table 1). At present there is controversy surrounding the use of neuromodulation for pelvic pain syndromes. Since the majority of practitioners who see CPP in clinical practice do not have primary training in neuromodulation, the range of techniques and approaches appears overwhelming, and it may be difficult to decide when neuromodulation is appropriate and which technique to choose. The aim of this overview is to summarize the evidence for the use of different neuromodulatory approaches in the management of chronic pelvic, bladder and prostatic, groin, and genital pain syndromes.

Table 1. Summary of types of neuromodulation technique and application for pelvic pain.

Neuromodulation Technique	Description	Indications	Advantages	Disadvantages	References
Percutaneous posterior tibial nerve stimulation	Placement of a fine needle into the posterior tibial nerve approximately 5 cm cephalad to the medial malleolus	Bladder pain syndrome (BPS), Chronic pelvic pain/Chronic prostatitis (CPP/CP)	Minimally invasive, low-risk, easier to perform, relatively cost-effective, no long-term follow-up needed	Need for patients to attend clinic weekly for 12 weeks to complete treatment. Minor side effects including mild pain and bleeding.	[37–45]
Implantable peripheral nerve stimulation devices	Implantation of insulated wire connected to implantable pulse generator to stimulate selected nerve (e.g., pudendal nerve)	Pudendal nerve (BPS, CPP/CP, pudendal neuralgia) genitofemoral, ilioinguinal, iliohypogastric (groin/genital pain)	Good specificity of effect	Requires technical skill, risk of infection, lead migration, and need for long-term follow-up	[46–50]
Sacral neuromodulation	Stimulation of sacral nerve roots by an electric current via an implanted insulated lead wire placed usually along the S3 sacral nerve root	CPP/CP, BPS, groin pain	Relatively widely used, so good evidence base to guide treatment.	Infection, lead migration or malfunction of the pulse generator or pain at the pulse generator site. Challenges in electrode placement.	[17,51–57]
Dorsal root ganglion stimulation	Implantation of an electrode connected to implantable pulse generator over the dorsal root ganglion	Pelvic girdle pain, groin pain	Long-term analgesic effects and specific anatomical targeting of the pain relief, as well as fewer changes in analgesic effect with changes in body posture	Requires technical skill, risk of infection, lead migration, and need for long-term follow up. Fewer large well-conducted trials into DRG stimulation for pelvic pain due to the fact that it is relatively new as a technique for this indication	[58–60]
Spinal cord stimulation	Implantation of an electrode over the dorsal spinal cord in the epidural space	CPP/CP, particularly pudendal neuralgia	Good efficacy in limited number of reported cases	Small number of studies carried out.	[61–66]
Motor cortex stimulation	Stimulation of motor cortex by placement of electrode in epidural space	CPP	May be an option in patients for whom peripheral or spinal neuromodulation was unsuccessful or contraindicated	Limited evidence	[67]
Deep brain stimulation	Stimulation of specific intracranial target by stereotactically placed electrodes	N/A	May be an option in patients for whom peripheral or spinal neuromodulation was unsuccessful or contraindicated	Limited evidence	[68]

2. Neuromodulation for Pelvic-Pain Syndromes

2.1. Peripheral Nerve Neuromodulation

2.1.1. Sacral Neuromodulation

Sacral neuromodulation (SNM) is a neuromodulatory technique whereby the sacral nerve is stimulated by an electric current via an implanted insulated lead wire placed along the sacral nerve roots, usually at the level of the S3 root. Though the precise mechanism of action of neuromodulation in relieving bladder and pelvic pain is not well understood, many publications suggest an effect on the modulation of spinal cord reflexes and brain networks, thus affecting bladder function [69,70]. First, patients are subject to a neuromodulation trial. Those who experience reduced pain with stimulation (classified as "responders"; often defined as at least 50% reduction in pain) are then allowed to progress on to implantation of a permanent implantable pulse generator (IPG), which is sited in a subcutaneous pocket in the lower quadrant of the abdomen or upper buttock and provides electrical stimulation [71]. The trial period is important as it helps prevent placement of an expensive permanent device, with its associated side effects, into a patient who may subsequently not respond to the therapy.

SNM was first approved for use for overactive-bladder syndrome and nonobstructive urinary retention. The initial use of SNM for pelvic pain came about following reports that pain symptoms improved in patients receiving SNM for a primary complaint of urinary symptoms, such as frequency and urgency [17]. It has since been trialed off-label as a treatment for pain in chronic pelvic pain, including some cohorts with patients with a variety of pelvic pain complaints, and other studies recruiting a narrower symptom range, such as bladder pain syndrome alone. In a prospective multicenter study, Martelluci et al. (2011) [51] reported results of SNM in 27 patients with multietiological medication-resistant pelvic pain. Trial stimulation was carried out initially, with an implantation rate of 59%, and significant improvements in visual analogue scale (VAS) pain score in those patients who were implanted, both at six months and subsequent follow up (mean follow-up duration was 37 months; mean preoperative VAS was 8.1, and mean VAS at six-month follow-up was 2.1). The study team attempted to evaluate the differences between patients who had a successful trial of SNM and those who did not. They noted that all the patients who had reported some benefit from gabapentin or pregabalin ($n = 9$) went on to have definitive SNM implantation, and that all patients who reported pain following a hysterectomy had permanent implantation ($n = 4$). Furthermore, all patients who correlated pain onset with previous surgery with stapler did not experience benefit during the stimulation trial and did not go ahead with SNM implantation ($n = 5$).

Sokal et al. (2015) [52] and Seigel et al. (2001) [17] both describe small single-center case series with good initial pain relief. The study by Seigel et al. (2001) is a prospective nonrandomized study, recruiting patients with intractable pelvic and/or urogenital pain. Results were reported from 10 patients (nine female and one male; median age 48 years; median pain duration 3 years) who all experienced >40% improvement in pain symptoms with test stimulation on an outpatient basis, and subsequently had the system implanted. Although no statistical analysis was reported, in 9 out of the 10 patients, the worst pain decreased (average decrease from 4.7 to 2.2 at long-term follow-up), and there were also improvements in other measures, such as the number of hours of worst pain and the rate of pain. However, among the 10 patients there were 27 complications reported, including local wound complications ($n = 6$), change in the location of the pain ($n = 4$), IPG site pain ($n = 4$), and implant infection ($n = 1$). Sokal et al (2015) [52] report outcomes of a prospective single-center study that recruited nine female patients with chronic pelvic pain (four as a result of failed back-surgery syndrome, and five as a result of idiopathic chronic regional pain syndrome). There was a statistically significant reduction in pain VAS at the six-month follow-up (median VAS 3 from preoperative level of 9), but reduction in efficacy at 12 months (median VAS 6), and higher than expected rate of complications, including infection and lead migration. In a mixed multicenter cohort of patients with urinary symptoms and/or perineal pain, Everaert et al. (2000) [53] also found good initial response rates to SNM (85%), which declined somewhat at a longer-term follow-up (70%). They also found that there was a significant relationship between psychiatric comorbidity and reported outcome, highlighting this as an important variable to further study in the context of SNM for chronic pelvic pain.

SNM has also been used with good effect in two patients with intractable pelvic pain following cauda equina syndrome, and had beneficial effects on the urinary symptoms experienced by these patients (Kim et al. (2010)) [54]. In bladder-pain syndrome specifically, we reviewed three prospective studies, including a total of 137 participants, which evaluated the efficacy of sacral neuromodulation in the management of BPS. Since its introduction for the management of bladder pain, SNM has been shown to have both subjective and objective improvements in symptomatology in patients with BPS with good long-term results [55–57]. Results include an increase in mean voided volume, reduced pain perception, reduced urinary frequency and nocturia, and an improvement in quality of life.

Overall, these initial trials of SNM for chronic pelvic pain suggest that it is effective for selected patients, including the BPS population, although current data relate predominantly to female rather than male patients and randomized controlled trials are difficult to identify; most studies are prospective observational trials involving patients with medication-refractory pelvic pain. Interestingly, the reported side-effect profile is relatively high, at about 3%, the most common of which being infection, lead migration, or malfunction of the pulse generator [36,72]. Other disadvantages include the fact that the procedure is expensive, which limits its use in routine clinical practice [69]. In addition, placement of the device requires specific surgical skills, which necessitates referral to a specialist and the associated waiting list. Patients also require life-long follow-ups if deemed suitable for management with sacral neuromodulation.

However, even taking disadvantages into account, the benefits afforded to medication-refractory patients by SNM strongly imply that this procedure should always be considered prior to major surgical intervention, such as augmentation procedures, urinary diversion, or cystectomy, for the purposes of pain control. Though the revision rate is high, with 49% of implanted devices requiring revision over an average follow-up of 38 months, this procedure is completely reversible, with minimal side effects of revision [73].

2.1.2. Posterior Tibial Nerve Stimulation

Posterior tibial nerve stimulation (PTNS) was first described by McGuire et al. (1983) for the treatment of detrusor instability, although the original series included five patients with interstitial cystitis, of whom four improved with stimulation [74]. Early studies investigated its efficacy as a treatment for lower urinary tract symptoms [75–77]. However, it was observed that patients reported a concomitant improvement in levels of pain resulting in trials of PTNS as a treatment for pelvic pain. PTNS is performed in the outpatient setting on a weekly basis for a period of 12 weeks, and delivers electrical stimulation to the sacral micturition center via the sacral nerve plexus S2–4. The mechanism of action is thought to involve inhibition or modulation of the C-fiber and Aδ-afferent responses from the bladder. It is a minimally invasive procedure, which involves the placement of a fine needle into the posterior tibial nerve, approximately 5 cm cephalad to the medial malleolus [78].

The major randomized controlled trial of PTNS for chronic pelvic pain was described by Kabay et al. (2009) [37]. Of 89 patients recruited to the trial, 45 were randomised to PTNS and 44 to a sham-treatment group. Patients were randomized from a multietiological pelvic-pain group (including subjects with pain in the bladder, groin, genitals, lower abodomen, perineum, and/or perianal area), and all were male. The mean age was 37.9 years in the treatment group and 38.5 years in the sham-stimulation group, and the mean disease duration was 4.5 years (treatment group) and 3.8 years (sham-stimulation group). Stimulation was carried out in 200 µs pulses, with an amplitude range of 1–10 mA. Significant improvements in VAS for pain, urgency, and National Institutes of Health Chronic Prostatitis Symptom Index (NIH-CPSI) scores were achieved in the PTNS group at follow-up (follow-up assessments were completed at 12 weeks, immediately after the treatment) but not the sham-controlled group. The mean VAS for pain improved from 7.6 ± 0.8 to 4.3 ± 0.6 in the PTNS group, with no change in the control group (scores were 7.4 ± 0.9 before sham treatment and 7.2 ± 0.4 after sham treatment). Overall, 40% of patients in the PTNS arm of the study achieved >50% improvement in VAS. This study demonstrates moderate and statistically significant benefits of PTNS over sham stimulation in the selected population, although outcomes at longer-term follow-up are not reported. Other studies, albeit smaller in recruitment numbers, have also reported moderately positive outcomes for PTNS in chronic pelvic pain [38–40], using similar stimulation parameters to those described by Kabay et al. [37]. Two other randomized controlled trials, both including only female patients with mixed etiology chronic pelvic pain, reported significant improvements in VAS for the PTNS group, but not the control group at a 12-week follow-up [39,40], although the statistical significance was not maintained at the six-month follow-up [39]. However, marked improvements in pain have not been reported by all studies; for example, in a prospective cohort study, enrolling male and female patients with mixed etiology pelvic pain, and a mean age of 51 years, only 21% of patients reported an improvement in VAS of >50% and, although the improvement in VAS for the group was statistically significant, the magnitude of change in VAS was small (from 6.5 at baseline to 5.4 after treatment) [41]. The authors note that improvement was better in patients with certain pain distributions (e.g., perineal, perianal, and vaginal) and that increasing the frequency of stimulation from once per week to more regularly might result in better outcomes based on findings from the overactive-bladder literature [76].

PTNS has also been tested as an experimental treatment for BPS. However, there are conflicting reports on its efficacy. A study by Congregado and colleagues reported significant improvements in all irritative lower urinary tract symptoms and hypogastric pain after 10 weeks of treatments with PTNS [42]. The study was a prospective observational follow-up study in 51 female patients with lower-urinary-tract irritative symptoms who had experienced no prior response to anticholinergic medications. All patients appeared to report hypogastric pain prior to treatment, but only 33% reported hypogastric pain at follow-up (mean follow-up duration was 21 months). Unfortunately, it is not clear in the paper how hypogastric pain was evaluated [76], and it appears that patients were recruited on the basis of their irritative symptoms and not primarily bladder pain. In contrast with these positive findings, Zhao et al. (2008) reported no significant change in the VAS pain score of BPS patients after a 10-week trial of PTNS in an open prospective trial, though significant improvement in bladder volume was noted, along with complete pain resolution in a single subject and statistically significant improvements in other secondary measures, such as the Interstitial Cystitis Problem Index and the O'Leary/Sant Interstitial Cystitis Problem Index [44]. Subsequently, this result was echoed in a trial by Regab and colleagues where they reported no effect on BPS symptoms following intermittent PTNS after 0, six, and 12 weeks of treatment [45]. Using a slightly different treatment approach, Baykal et al. found that PTNS when used in combination with glycosaminoglycan replacement therapy (intravesical heparin) was effective in improving pain scores and bladder capacity in refractory BPS patients (10 female, two male) who had failed "more than one classical therapy" [43]. There was no control arm in this study to compare the effects of PTNS + intravesical heparin with intravesical heparin alone; however, positive results suggest that this should be investigated further.

Overall, the main advantage of PTNS is that it is minimally invasive, with only mild side effects (predominantly pain at the insertion site and mild bleeding or bruising [38,40,44]) compared with other types of neuromodulation, and, as a result of that, patients tend to find it acceptable [44]. There appears to be a moderate benefit of PTNS for pelvic pain in medication-refractory patients [38–40,42,43], but the benefits may tail off over time, since the stimulator is not permanently implanted, unlike SNM. Reduction in efficacy at long-term follow-up was found by Istek et al. [34], and it is clear that more long-term follow-up studies are needed to investigate this further. Large prospective randomized controlled trials that are able to compare the effect of PTNS with sham stimulation and also identify phenotypes within the pelvic-pain spectrum that respond more favorably to stimulation would be an important next step. Further trials of combination therapy (for example, glycosaminoglycan replacement + PTNS) may also be of benefit in more fully exploring the role of PTNS in pelvic pain.

2.1.3. Pudendal Nerve Stimulation

Pudendal nerve stimulation can be successful for pelvic pain when the pain is identified as being perineal in nature, and if the pain is associated with features of pudendal neuralgia. As with sacral neuromodulation, the technique can be carried out as a two-stage procedure, with a lead positioned at the pudendal nerve for test stimulation, and connected to an implantable pulse generator if the test stimulation proves successful. Peters et al. (2015) [79] conducted a retrospective review in which 19 patients who had undergone pudendal neuromodulation at a single center for pudendal neuralgia were sent questionnaires to evaluate outcome. All patients had had some improvement in pain at the time of implantation. Only 10 out of 19 patients returned the questionnaires; of these, seven reported some improvement (four reported slight improvement, one reported moderate improvement, and two reported marked improvement). However, pain medications received more favorable assessments, with six out of 10 patients describing a marked improvement. In a case series of three patients, Carmel et al. (2009) [26] reported more favorable outcomes, with one patient pain-free at two-year follow-up, and two patients reporting 80% pain relief. However, numbers are small and further studies are needed to strengthen the evidence for this treatment strategy. In cohorts of patients with BPS, pudendal neuromodulation has been shown in several studies and case reports to be effective in alleviating pain, especially in patients who have failed management with sacral neuromodulation. However, this method is new, has limited evidence, and is therefore not routinely practiced. We reviewed three studies with a total number of 102 subjects that evaluated the role of pudendal nerve stimulation in the management of patients with BPS. The first, a retrospective study on 84 patients concluded that pudendal neuromodulation could be recommended in patients who are refractory to sacral neuromodulation: 93% of patients who had previously failed sacral neuromodulation responded to pudendal stimulation [80]. When compared to sacral nerve stimulation in a blinded randomized trial design study, this approach was reported to lead to significantly greater reduction in bladder pain and irritative urinary symptoms in complex BPS patients [80,81]. Finally, pudendal neuromodulation was described in a case report, in combination with sacral neuromodulation, to produce excellent results for the treatment of complex pelvic neuropathy [82]. The pudendal nerve may thus play a more important role in the management of BPS than is currently recognized in daily clinical practice.

2.1.4. Stimulation of Other Peripheral Nerves

Although PTNS and sacral neuromodulation are by far the most common nerve stimulation techniques for chronic pelvic pain, followed by pudendal neuromodulation, neuromodulation of other peripheral nerves, including the genitofemoral, ilioinguinal, iliohypogastric, and vagus nerves, has been successfully performed in small numbers of patients for intractable inguinal pain [46–48]. Carayannopoulos et al. (2009) [46] published outcomes for two patients; the first had medication-refractory pain in the inguinal, genital, and thigh regions, which had temporarily responded to ilioinguinal nerve blocks and pulsed radiofrequency ablation of the ilioinguinal nerve, and the second had groin pain that was not completely relieved by medications. Patients reported 90% and 85%–95% pain alleviation seven days after implantation; however, longer-term follow-up data were not provided. A study by Shaw et al. (2016) [47] included six patients with chronic neuropathic inguinal and genital pain (four male, two female, mean pain duration 4.6 years). All patient had undergone trials of other therapies, including medication and nerve blocks, for the treatment of their pain. Five out of six patients had sustained benefit with stimulation at long-term follow-up (average follow-up duration was 22 months) and two patients had a VAS pain score of zero at that point. Testicular pain following hydrocele surgery has also been reported as responding well to stimulation of the cutaneous branch of the ilioinguinal, and the genital branch of the genitofemoral nerves in a case report by Rosendal et al. 2012 [49]. Pain intensity reduced from 9/10 to 2/10 at seven-month follow-up in this patient.

Stimulation of the vagus nerve has been attempted for the control of pelvic pain, based on evidence that the vagus nerve plays a role in visceral nociception. In a study of 15 female subjects, Napadow et al. 2012 [50] investigated the effect of respiratory-gated auricular vagal afferent nerve stimulation on pain relief in patients with chronic pelvic pain. The study used a randomized crossover design comparing a single session of respiratory-gated auricular vagal afferent nerve stimulation with a single session of auricular stimulation, which was nonvagal. They found that patients undergoing the respiratory-gated auricular vagal afferent nerve stimulation had significantly less anxiety than the nonvagal stimulation group, and that there was a trend towards reductions in evoked pain intensity and temporal summation of evoked pain in the respiratory-gated auricular vagal afferent nerve stimulation group.

2.2. Dorsal-Root-Ganglion Stimulation

The DRG is a collection of cell bodies of sensory neurons that is located bilaterally at each spinal level encased within the bony vertebral structure. As part of the anatomical pathway involved in pain transmission, electrical stimulation of the DRG has been explored as a treatment for chronic pain [59].

DRG stimulation, with the stimulating electrodes at L1 and L2 level, has been reported in a single case of intractable, medication-resistant pelvic girdle pain, with a 43% reduction in pain at six-month follow-up [60]. DRG stimulation has also been described for groin pain. Sensory input to DRGs at T11-L3 corresponds to the groin area. In a multicenter study of DRG stimulation for chronic pain, 10/10 patients with postherniorrhaphy pain had a successful stimulation trial, and the mean reduction in VAS score at follow-up was 76.8 ± 8.2% [58]. Larger prospective studies are awaited.

2.3. Spinal-Cord Stimulation

Spinal-cord stimulation (SCS) a common neurostimulation approach for the treatment of chronic pain, first reported early in the second half of the twentieth century [83], which involves surgical laminotomy and placement of electrodes in the epidural space between T9 and T11 for lower-limb pain. Its mechanism of action is thought to involve modulation of pain transmission in the spinal-cord dorsal horn, in addition to manipulating autonomic function and interacting with supraspinal pain-processing mechanisms. Although there is good evidence for its use in severe pain associated with failed back-surgery syndrome, chronic regional pain syndrome, and neuropathic pain, far less is known about its efficacy for visceral pain and pelvic pain. However, there are a small number of studies describing its use in this context [61–63]. Buffenoir et al. (2015) [61] report outcomes following SCS at the conus medullaris in a prospective dual-center study enrolling a total of 27 patients with pudendal neuralgia, recruited over a 13-month period. Twenty out of 27 patients were classified as 'responders' (>50% reduction of maximum pain or >50% increase of sitting time before pain onset). The estimated percentage improvement at long-term follow-up was 55% with a mean tripling of sitting time. Short-term complications included one infection and one suboptimal electrode fixation but no long-term complications were described. This technique has been recently replicated in a small case series with good results [63]. SCS at T7–9 levels for groin/inguinal pain has also been reported as beneficial in small groups of patients with postherniorrhaphy groin pain, [64,65]. However, SCS-induced parasthesias may not always fully cover the groin area, and in this case, a combination of SCS with peripheral field stimulation may be useful [66]. Despite these studies, the standard of evidence for SCS in the context of pelvic pain remains of a fairly low quality and further research is needed to define the appropriate context for this technique.

2.4. Brain Stimulation for Pelvic-Pain Syndromes: Existing Evidence and Future Directions

The peripheral afferent drive is known to be important in chronic pain, including chronic pelvic-pain syndromes. As a demonstration of this, intravesical installation of alkalinized lidocaine has been shown to have benefits for selected patients with bladder-pain syndrome [84], and is thought to have its effects by silencing the afferent pain drive [6]. However, in patients who do not effectively respond to such treatment, it is reasonable to assume that central mechanisms contribute substantially to their experience of pain. In such cases of chronic pelvic pain where central sensitization plays a key role in the development and maintenance of the chronic pain state, peripheral approaches to neuromodulation may fail to address the root cause of the problem, and a strategy of central neuromodulation may be more effective for symptom control. Indeed, a test of the contribution of the peripheral afferent drive, such as intravesical lidocaine, might become a tool for selecting candidates for central, supraspinal neuromodulation. Brain stimulation techniques for the control of pain include motor cortex stimulation (MCS), in which epidural electrodes are sited over the motor cortex, and deep brain stimulation (DBS) in which electrodes are implanted at targets within the brain itself, including the periaqueductal/periventricular grey area, the ventral posterolateral and ventral posteromedial thalamus, and the anterior cingulate cortex. In the case of chronic pelvic pain, MCS stimulation for pelvic and perineal pain has been described at case-report level to provide improvement in medication-refractory cases that have failed an alternative neuromodulation trial, or those for which peripheral or spinal neuromodulation is contraindicated [67]. Far more trial data is needed to determine if this should be considered routinely in refractory pelvic pain. Similarly, there may be a potential role for DBS in the control of pelvic pain. At the time of writing, there did not appear to be any studies describing individual outcomes following DBS for pelvic, groin, or genital pain, although DBS has been performed for these indications, with outcomes reported as averages within larger series, (e.g., Reference [68]), but in which it is not possible to identify specific outcomes for pelvic-pain patients.

3. Conclusions

3.1. What We Know

Chronic pelvic pain is a major area of unmet clinical need, with massive associated morbidity and health costs, and encompassing a wide range of different pain syndromes. The underlying pathophysiology is heterogeneous and likely to involve both peripheral and central mechanisms. Neuromodulation is an emerging option for patients with refractory pelvic pain, and both PTNS and SNM are recognized as potential treatments. Our conclusion is that peripheral neuromodulation, such as PTNS, SNM, or pudendal nerve stimulation, should be considered in patients whose pain is refractory to medication, particularly if they have shown some response to a nerve block. PTNS has a better side-effect profile than SNM, but its effects seem to be more short-lived. Other neuromodulation techniques, such as DRG stimulation and spinal cord stimulation, are still highly experimental but may also be considered in selected patients.

3.2. What We Do Not Know

There are many gaps in the current literature regarding neuromodulation for urogenital and chronic pelvic pain. Firstly, knowledge about underlying pathophysiological mechanisms of pain in chronic pelvic pain syndromes, particularly the role of peripheral and central mechanisms maintaining the pain state, is still incomplete. There is a shortage of large, randomized controlled trials of neuromodulation therapies for chronic pelvic pain, and it is therefore difficult to fully assess efficacy. Furthermore, most studies focus on female patients, and lack long-term follow-up, so the long-term effectiveness and relevance for male patients is not known. There are no direct comparisons between neuromodulation types, and little is understood about which subgroups and phenotypes might respond better to different types of neuromodulation. Finally, knowledge about the potential of spinal cord and brain stimulation for pelvic and urogenital pain is limited to case-report level only and further studies are needed.

3.3. Limitations of This Overview

This overview is limited in that it is not a formal systematic review. We have also only included work published in the English language, which may have limited the article's scope. Finally, we did not attempt to contact the authors of studies that included pelvic pain patients within larger series without specifically providing a breakdown of results for pelvic pain patients. Therefore, knowledge available from these studies, for example, the work on DBS for pelvic pain, was not accessed here.

Author Contributions: Conceptualization, H.R. and I.O.; Methodology, H.R. and I.O.; Writing—Original draft preparation, H.R. and I.O.; Writing—review and editing, H.R., I.O., and A.D.; Supervision, A.D.

Funding: This research received no external funding.

Acknowledgments: Many thanks to Christopher Roy for help with the figure design.

Conflicts of Interest: The authors declare no conflict of interest.

References

1. Kanai, A.; Andersson, K.E. Bladder Afferent Signaling: Recent Findings. *J. Urol.* **2010**, *183*, 1288–1295. [CrossRef] [PubMed]

2. Grundy, L.; Brierley, S.M. Cross-organ sensitization between the colon and bladder: To pee or not to pee? *Am. J. Physiol. Gastrointest. Liver Physiol.* **2018**, *314*. [CrossRef] [PubMed]

3. Panicker, J.N.; Manji, H. Neuromuscular Disorders. In *Pelvic Organ Dysfunction in Neurological Disease: Clinical Management and Rehabilitation*; Clare, J.F., Jalesh, N.P., Emmanuel, A., Eds.; Cambridge University Press: London, UK, 2010.

4. Graziottin, A.; Gambini, D. Anatomy and physiology of genital organs–women. *Handb. Clin. Neurol.* **2015**, *130*, 39–60. [PubMed]

5. Elkins, N.; Hunt, J.; Scott, K.M. Neurogenic pelvic pain. *Phys. Med. Rehabil. Clin. N. Am.* **2017**, *28*, 551–569. [CrossRef] [PubMed]

6. Gebhart, G.F.; Bielefeldt, K. Physiology of visceral pain. *Compr. Physiol.* **2016**, *6*, 1609–1633. [PubMed]

7. Inoue, K.; Tsuda, M. Microglia in neuropathic pain: Cellular and molecular mechanisms and therapeutic potential. *Nat. Rev. Neurosci.* **2018**, *19*, 138–152. [CrossRef] [PubMed]

8. Bliss, T.V.P.; Collingridge, G.L.; Kaang, B.K.; Zhuo, M. Synaptic plasticity in the anterior cingulate cortex in acute and chronic pain. *Nat. Rev. Neurosci.* **2016**, *17*, 485–496. [CrossRef] [PubMed]

9. Rana, N.; Drake, M.J.; Rinko, R.; Dawson, M.; Whitmore, K. The fundamentals of chronic pelvic pain assessment, based on international continence society recommendations. *Neurourol. Urodyn.* **2018**, *37*, S32–S38. [CrossRef]

10. Weissman, E.; Boothe, E.; Wadhwa, V.; Scott, K.; Chhabra, A. Magnetic resonance neurography of the pelvic nerves. *Semin. Ultrasound CT MR* **2017**, *38*, 269–278. [CrossRef] [PubMed]

11. Chen, A.; De, E.; Argoff, C. Small fiber polyneuropathy is prevalent in patients experiencing complex chronic pelvic pain. *Pain Med.* **2018**. [CrossRef] [PubMed]

12. Eller-Smith, O.C.; Nicol, A.L.; Christianson, J.A. potential mechanisms underlying centralized pain and emerging therapeutic interventions. *Front. Cell. Neurosci.* **2018**, *12*, 35. [CrossRef] [PubMed]

13. Levesque, A.; Riant, T.; Ploteau, S.; Rigaud, J.; Labat, J.J. Clinical criteria of central sensitization in chronic pelvic and perineal pain (convergences PP criteria): Elaboration of a clinical evaluation tool based on formal expert consensus. *Pain Med.* **2018**. [CrossRef] [PubMed]

14. Tam, J.; Loeb, C.; Grajower, D.; Kim, J.; Weissbart, S. Neuromodulation for chronic pelvic pain. *Curr. Urol. Rep.* **2018**, *19*, 32. [CrossRef] [PubMed]

15. Tutolo, M.; Ammirati, E.; Heesakkers, J.; Kessler, T.M.; Peters, K.M.; Rashid, T.; Sievert, K.D.; Spinelli, M.; Novara, G.; Van der Aa, F. Efficacy and safety of sacral and percutaneous tibial neuromodulation in non-neurogenic lower urinary tract dysfunction and chronic pelvic pain: A systematic review of the literature. *Eur. Urol.* **2018**, *73*, 406–418. [CrossRef] [PubMed]

16. Baranowski, A.P.; Lee, J.; Price, C.; Hughes, J. Pelvic pain: A pathway for care developed for both men and women by the British Pain Society. *Br. J. Anaesth.* **2014**, *112*, 452–459. [CrossRef] [PubMed]

17. Siegel, S.; Paszkiewicz, E.; Kirkpatrick, C.; Hinkel, B.; Oleson, K. Sacral nerve stimulation in patients with chronic intractable pelvic pain. *J. Urol.* **2001**, *166*, 1742–1745. [CrossRef]

18. Parasar, P.; Ozcan, P.; Terry, K.L. Endometriosis: Epidemiology, diagnosis and clinical management. *Curr. Obstetrics Gynecol. Rep.* **2017**, *6*, 34–41. [CrossRef] [PubMed]

19. Abrams, P.; Cardozo, L.; Fall, M.; Griffiths, D.; Rosier, P.; Ulmsten, U.; van Kerrebroeck, P.; Victor, A.; Wein, A. The standardisation of terminology of lower urinary tract function: Report from the Standardisation Sub-committee of the International Continence Society. *Neurourol. Urodyn.* **2002**, *21*, 167–178. [CrossRef] [PubMed]

20. Leppilahti, M.; Tammela, T.L.; Huhtala, H.; Auvinen, A. Prevalence of symptoms related to interstitial cystitis in women: A population-based study in Finland. *J. Urol.* **2002**, *168*, 139–143. [CrossRef]

21. Rosenberg, M.T.; Hazzard, M. Prevalence of interstitial cystitis symptoms in women: A population-based study in the primary care office. *J. Urol.* **2005**, *174*, 2231–2234. [CrossRef] [PubMed]

22. Graham, E.; Chai, T.C. Dysfunction of bladder urothelium and bladder urothelial cells in interstitial cystitis. *Curr. Urol. Rep.* **2006**, *7*, 440–446. [CrossRef] [PubMed]

23. Keay, S.K.; Birder, L.A.; Chai, T.C. Evidence for bladder urothelial pathophysiology in functional bladder disorders. *BioMed Res. Int.* **2014**. [CrossRef] [PubMed]

24. Cheong, Y.; Saran, M.; Hounslow, J.W.; Reading, I.C. Are pelvic adhesions associated with pain, physical, emotional and functional characteristics of women presenting with chronic pelvic pain? A cluster analysis. *BMC Womens Health* **2018**, *18*, 11. [CrossRef] [PubMed]

25. Hunter, C.W.; Stovall, B.; Chen, G.; Carlson, J.; Levy, R. Anatomy, pathophysiology and interventional therapies for chronic pelvic pain: A review. *Pain Physician* **2018**, *21*, 147–167. [PubMed]

26. Carmel, M.; Lebel, M.; Tu le, M. Pudendal nerve neuromodulation with neurophysiology guidance: A potential treatment option for refractory chronic pelvi-perineal pain. *Int. Urogynecol. J.* **2010**, *21*, 613–616. [CrossRef] [PubMed]

27. Manikandan, R.; Srirangam, S.J.; Pearson, E.; Collins, G.N. Early and late morbidity after vasectomy: A comparison of chronic scrotal pain at 1 and 10 years. *BJU Int.* **2004**, *93*, 571–574. [CrossRef] [PubMed]

28. Christofferson, M.; Barnard, J.; Montoya, T.I. Clitoral pain following retropubic midurethral sling placement. *Sex. Med.* **2015**, *3*, 346–348. [CrossRef] [PubMed]

29. Kehlet, H.; Jensen, T.S.; Woolf, C.J. Persistent postsurgical pain: risk factors and prevention. *Lancet* **2006**, *367*, 1618–1625. [CrossRef]

30. Poobalan, A.S.; Bruce, J.; Smith, W.C.; King, P.M.; Krukowski, Z.H.; Chambers, W.A. A review of chronic pain after inguinal herniorrhaphy. *Clin. J. Pain* **2003**, *19*, 48–54. [CrossRef] [PubMed]

31. Aasvang, E.; Kehlet, H. Surgical management of chronic pain after inguinal hernia repair. *Br. J. Surg.* **2005**, *92*, 795–801. [CrossRef] [PubMed]

32. Chiarioni, G.; Asteria, C.; Whitehead, W.E. Chronic proctalgia and chronic pelvic pain syndromes: Etiologic insights and treatment options. *World J. Gastroenterol.* **2011**, *17*, 4447–4455. [CrossRef] [PubMed]

33. Fall, M.; Baranoski, A.; Elneil, S.; Engeler, D.; Hughes, J.; Messelink, E.J.; Oberpenning, F.; de Williams, A.C. EAU guidelines on chronic pelvic pain. *Eur. Urol.* **2010**, *57*, 35–48. [CrossRef] [PubMed]

34. Cheong, Y.C.; Smotra, G.; Williams, A.C.D.C. Non-surgical interventions for the management of chronic pelvic pain. *Cochrane Database Syst. Rev.* **2014**. [CrossRef] [PubMed]

35. Nickel, J.C. The three as of chronic prostatitis therapy: Antibiotics, alpha-blockers and anti-inflammatories. What is the evidence? *BJU Int.* **2004**, *94*, 1230–1233. [CrossRef] [PubMed]

36. Wang, J.; Chen, Y.; Chen, J.; Zhang, G.; Wu, P. Sacral Neuromodulation for refractory bladder pain syndrome/interstitial cystitis: A global systematic review and meta-analysis. *Sci. Rep.* **2017**, *7*, 11031. [CrossRef] [PubMed]

37. Kabay, S.; Kabay, S.C.; Yucel, M.; Ozden, H. Efficiency of posterior tibial nerve stimulation in category IIIB chronic prostatitis/chronic pelvic pain: A Sham-Controlled Comparative Study. *Urol. Int.* **2009**, *83*, 33–38. [CrossRef] [PubMed]

38. Kim, S.W.; Paick, J.S.; Ku, J.H. Percutaneous posterior tibial nerve stimulation in patients with chronic pelvic pain: A preliminary study. *Urol. Int.* **2007**, *78*, 58–62. [CrossRef] [PubMed]

39. Istek, A.; Gungor Ugurlucan, F.; Yasa, C.; Gokyildiz, S.; Yalcin, O. Randomized trial of long-term effects of percutaneous tibial nerve stimulation on chronic pelvic pain. *Arch. Gynecol. Obstet.* **2014**, *290*, 291–298. [CrossRef] [PubMed]

40. Gokyildiz, S.; Kizilkaya Beji, N.; Yalcin, O.; Istek, A. Effects of percutaneous tibial nerve stimulation therapy on chronic pelvic pain. *Gynecol. Obstet. Investig.* **2012**, *73*, 99–105. [CrossRef] [PubMed]

41. Van Balken, M.R.; Vandoninck, V.; Messelink, B.J.; Vergunst, H.; Heesakkers, J.P.; Debruyne, F.M.; Bemelmans, B.L. Percutaneous tibial nerve stimulation as neuromodulative treatment of chronic pelvic pain. *Eur. Urol.* **2003**, *43*, 158–163. [CrossRef]

42. Congregado Ruiz, B.; Pena Outeirino, X.M.; Campoy Martinez, P.; Leon Duenas, E.; Leal Lopez, A. Peripheral afferent nerve stimulation for treatment of lower urinary tract irritative symptoms. *Eur. Urol.* **2004**, *45*, 65–69.

43. Baykal, K.; Senkul, T.; Sen, B.; Karademir, K.; Adayener, C.; Erden, D. Intravesical heparin and peripheral neuromodulation on interstitial cystitis. *Urol. Int.* **2005**, *74*, 361–364. [CrossRef] [PubMed]

44. Zhao, J.; Bai, J.; Zhou, Y.; Qi, G.; Du, L. Posterior tibial nerve stimulation twice a week in patients with interstitial cystitis. *Urology* **2008**, *71*, 1080–1084. [CrossRef] [PubMed]

45. Ragab, M.M.; Tawfik, A.M.; Abo El-enen, M.; Elnady, M.; El-Gamal, O.M.; El-Kordy, M.; Gameel, T.; Rasheed, M. Evaluation of percutaneous tibial nerve stimulation for treatment of refractory painful bladder syndrome. *Urology* **2015**, *86*, 707–711. [CrossRef] [PubMed]

46. Carayannopoulos, A.; Beasley, R.; Sites, B. Facilitation of percutaneous trial lead placement with ultrasound guidance for peripheral nerve stimulation trial of ilioinguinal neuralgia: A technical note. *Neuromodulation* **2009**, *12*, 296–301. [CrossRef] [PubMed]

47. Shaw, A.; Sharma, M.; Zibly, Z.; Ikeda, D.; Deogaonkar, M. Sandwich technique, peripheral nerve stimulation, peripheral field stimulation and hybrid stimulation for inguinal region and genital pain. *Br. J. Neurosurg.* **2016**, *30*, 631–636. [CrossRef] [PubMed]

48. Al Tamimi, M.; Davids, H.R.; Barolat, G.; Krutsch, J.; Ford, T. Subcutaneous peripheral nerve stimulation treatment for chronic pelvic pain. *Neuromodulation* **2008**, *11*, 4. [CrossRef] [PubMed]

49. Rosendal, F.; Moir, L.; de Pennington, N.; Green, A.L.; Aziz, T.Z. Successful treatment of testicular pain with peripheral nerve stimulation of the cutaneous branch of the ilioinguinal and genital branch of the genitofemoral nerves. *Neuromodulation* **2013**, *16*, 121–124. [CrossRef] [PubMed]

50. Napadow, V.; Edwards, R.R.; Cahalan, C.M.; Mensing, G.; Greenbaum, S.; Valovska, A.; Li, A.; Kim, J.; Maeda, Y.; Park, K.; et al. Evoked Pain Analgesia in Chronic Pelvic Pain Patients using Respiratory-Gated Auricular Vagal Afferent Nerve Stimulation. *Pain Med.* **2012**, *13*, 777–789. [CrossRef] [PubMed]

51. Martellucci, J.; Naldini, G.; Carriero, A. Sacral nerve modulation in the treatment of chronic pelvic pain. *Int. J. Colorectal Dis.* **2012**, *27*, 921–926. [CrossRef] [PubMed]

52. Sokal, P.; Zielinski, P.; Harat, M. Sacral roots stimulation in chronic pelvic pain. *Neurol. Neurochir. Pol.* **2015**, *49*, 307–312. [CrossRef] [PubMed]

53. Everaert, K.; De Ridder, D.; Baert, L.; Oosterlinck, W.; Wyndaele, J.J. Patient satisfaction and complications following sacral nerve stimulation for urinary retention, urge incontinence and perineal pain: A multicenter evaluation. *Int. Urogynecol. J. Pelvic Floor Dysfunct.* **2000**, *11*, 231–235. [CrossRef] [PubMed]

54. Kim, J.H.; Hong, J.C.; Kim, M.S.; Kim, S.H. Sacral nerve stimulation for treatment of intractable pain associated with cauda equina syndrome. *J. Korean Neurosurg. Soc.* **2010**, *47*, 473–476. [CrossRef] [PubMed]

55. Al-Zahrani, A.A.; Elzayat, E.A.; Gajaweski, J.B. Long-term outcome and surgical intervewntions after sacral neuromodulation implant for lower urinary tract symptoms: 14-year experience at 1 centre. *J. Urol.* **2011**, *185*, 981–986. [CrossRef] [PubMed]

56. Gajewski, J.B.; Al-Zahrani, A.A. The long-term efficacy of sacral neuromodulation in the management of intractable cases of bladder pain syndrome: 14 Years of experience in one centre. *BJU Int.* **2011**, *107*, 1258–1264. [CrossRef] [PubMed]

57. Ghazwani, Y.Q.; Elkelini, M.S.; Hassouna, M.M. Efficacy of sacral neuromodulation in treatment of bladder pain syndrome: Long-term follow-up. *Neurourol. Urodyn.* **2011**, *30*, 1271–1275. [CrossRef] [PubMed]

58. Schu, S.; Gulve, A.; ElDabe, S.; Baranidharan, G.; Wolf, K.; Demmel, W.; Rasche, D.; Sharma, M.; Klase, D.; Jahnichen, G.; et al. Spinal cord stimulation of the dorsal root ganglion for groin pain—A retrospective review. *Pain Pract.* **2015**, *15*, 293–299. [CrossRef] [PubMed]

59. Liem, L.; Russo, M.; Huygen, F.J.P.M.; van Buyten, J.P.; Smet, I.; Verrils, P.; Cousins, M.; Brooker, C.; Levy, R.; Deer, T.; Kramer, J. A Multicenter, Prospective trial to assess the safety and performance of the spinal modulation dorsal root ganglion neurostimulator system in the treatment of chronic pain. *Neuromodulation* **2013**, *16*, 471–482. [CrossRef] [PubMed]

60. Rowland, D.C.; Wright, D.; Moir, L.; FitzGerald, J.J.; Green, A.L. Successful treatment of pelvic girdle pain with dorsal root ganglion stimulation. *Br. J. Neurosurg.* **2016**, *30*, 685–686. [CrossRef] [PubMed]

61. Buffenoir, K.; Rioult, B.; Hamel, O.; Labat, J.J.; Riant, T.; Robert, R. Spinal cord stimulation of the conus medullaris for refractory pudendal neuralgia: A prospective study of 27 consecutive cases. *Neurourol. Urodyn.* **2015**, *34*, 177–182. [CrossRef] [PubMed]

62. Kapural, L.N.S.; Janicki, T.I.; Mekhail, N. Spinal cord stimulation is an effective treatment for the chronic intractable visceral pelvic pain. *Pain Med.* **2006**, *7*, 440–443. [CrossRef] [PubMed]

63. Simopoulos, T.Y.R.; Gill, J.S. Treatment of chronic refractory neuropathic pelvic pain with high frequency 10 kilohertz spinal cord stimulation. *Pain Pract.* **2018**, *18*, 805–809. [CrossRef] [PubMed]

64. Elias, M. Spinal cord stimulation for post-herniorrhaphy pain. *Neuromodulation* **2000**, *3*, 155–157. [CrossRef] [PubMed]

65. Yakovlev, A.E.; Resch, B.E. Treatment of intractable abdominal pain patient with Bannayan-Riley-Ruvalcaba syndrome using spinal cord stimulation. *WMJ* **2009**, *108*, 323–326. [PubMed]

66. Lepski, G.; Vahedi, P.; Tatagiba, M.S.; Morgalla, M. Combined spinal cord and peripheral nerve field stimulation for persistent post-herniorrhaphy pain. *Neuromodulation* **2013**, *16*, 84–88. [CrossRef] [PubMed]

67. Louppe, J.M.; Nguyen, J.P.; Robert, R.; Buffenoir, K.; de Chauvigny, E.; Riant, T.; Pereon, Y.; Labat, J.J.; Nizard, J. Motor cortex stimulation in refractory pelvic and perineal pain: Report of two successful cases. *Neurourol. Urodyn.* **2013**, *32*, 53–57. [CrossRef] [PubMed]

68. Boccard, S.G.; Pereira, E.A.; Moir, L.; Aziz, T.Z.; Green, A.L. Long-term outcomes of deep brain stimulation for neuropathic pain. *Neurosurgery* **2013**, *72*, 221–230. [CrossRef] [PubMed]

69. Abello, A.; Das, A.K. Electrical neuromodulation in the managemnt of lower urinary tract dysfunction: Evidence, experience and future prospects. *Ther. Adv. Urol.* **2018**, *10*, 165–173. [CrossRef] [PubMed]

70. Kessler, T.M.; de Wachter, S. Neuromodulation of lower urinary tract dysfunction. *Urol. A* **2017**, *56*, 1591–1596. [CrossRef] [PubMed]

71. Lucas, M.G.; Bedretdinova, D.; Bosch, J.L.H.R.; Burkhard, F.; Cruz, F.; Nambiar, A.K.; Nilsson, C.G.; de Ridder, D.J.M.K.; Tubaro, A.; Pickard, R.S. Guidelines on urinary incontinence. *Eur. Assoc. Urol.* **2012**, 210 236.

72. Zabihi, N.; Mourtzinos, A.; Maher, M.G.; Raz, S.; Rodriguez, L.V. Short-term results of bilateral S2-S4 sacral neuromodulation for the treatment of refractory interstitial cystitis, painful bladder syndrome, and chronic pelvic pain. *Int. Urogynecol. J. Pelvic Floor Dysfunct.* **2008**, *19*, 553–557. [CrossRef] [PubMed]

73. Donon, L.; Robert, G.; Ballanger, P. Sacral neuromodulation: results of a monocentric study of 93 patients. *Prog. Urol.* **2014**, *24*, 1120–1131. [CrossRef] [PubMed]

74. McGuire, E.J.; Zhang, S.C.; Horwinski, E.R.; Lytton, B. Treatment of motor and sensory detrusor instability by electrical stimulation. *J. Urol.* **1983**, *129*, 78–79. [CrossRef]

75. Stoller, M.L. Afferent nerve stimulation for pelvic floor dysfunction. *Eur. Urol.* **1999**, *35*, 16.

76. Klingler, H.C.; Pycha, A.; Schmidbauer, J.; Marberger, M. Use of peripheral neuromodulation of the S3 region for treatment of detrusor overactivity: A urodynamic-based study. *Urology* **2000**, *56*, 766–771. [CrossRef]

77. Govier, F.E.; Litwiller, S.; Nitti, V.; Kreder, K.J., Jr.; Rosenblatt, P. Percutaneous afferent neuromodulation for the refractory overactive bladder: Result of a multicenter study. *J. Urol.* **2001**, *166*, 914–918.

78. Gupta, P.; Ehlert, M.J.; Siris, L.T.; Peters, K.M. Percutaneous tibial nerve stimulation and sacral neuromodualtion: An update. *Curr. Urol. Rep.* **2015**, *16*, 4. [CrossRef] [PubMed]

79. Peters, K.M.; Killinger, K.A.; Jaeger, C.; Chen, C. Pilot study exploring chronic pudendal neuromodulation as a treatment option for pain associated with pudendal neuralgia. *Low Urin. Tract Symptoms* **2015**, *7*, 138–142. [CrossRef] [PubMed]

80. Peters, K.M.; Killinger, K.A.; Boguslawski, B.M.; Boura, J.A. Chronic pudendal neuromodulation: Expanding available treatment options for refractory urologic symptoms. *Neurourol. Urodyn.* **2010**, *29*, 1267–1271. [CrossRef] [PubMed]

81. Peters, K.M.; Feber, K.M.; Bennett, R.C. A prospective, single-blind, randomized crossover trial of sacral vs pudendal nerve stimulation for interstitial cystitis. *BJU Int.* **2007**, *100*, 835–839. [CrossRef] [PubMed]

82. Armstrong, G.L.; Vancaillie, T.G. Combined site-specific sacral neuromodulation and pudendal nerve release surgery in a patient with interstitial cystitis and persistent arousal. *BMJ Case Rep.* **2016**. [CrossRef] [PubMed]

83. Shealy, C.N.; Mortimer, J.T.; Reswick, J.B. Electrical inhibition of pain by stimulation of the dorsal columns: Preliminary clinical report. *Anesth. Analg.* **1967**, *46*, 489–491. [CrossRef] [PubMed]
84. Nickel, J.C.; Moldwin, R.; Lee, S.; Davis, E.L.; Henry, R.A.; Wyllie, M.G. Intravesical alkalinized lidocaine (PSD597) offers sustained relief from symptoms of interstitial cystitis and painful bladder syndrome. *BJU Int.* **2009**. [CrossRef] [PubMed]

Review

The Current State of Deep Brain Stimulation for Chronic Pain and Its Context in Other Forms of Neuromodulation

Sarah Marie Farrell [1], Alexander Green [1,2] and Tipu Aziz [1,2,*]

1   Nuffield Department of Surgical Sciences, John Radcliffe Hospital, University of Oxford, Oxford OX3 9DU, UK; sarah.farrell@medsci.ox.ac.uk (S.M.F.); alex.green@nds.ox.ac.uk (A.G.)
2   Nuffield Department of Clinical Neurosciences, John Radcliffe Hospital, University of Oxford, Oxford OX3 9DU, UK
*   Correspondence: tipu.aziz@nds.ox.ac.uk; Tel.: +44-18-6522-7645

Received: 17 July 2018; Accepted: 13 August 2018; Published: 20 August 2018

Abstract: Chronic intractable pain is debilitating for those touched, affecting 5% of the population. Deep brain stimulation (DBS) has fallen out of favour as the centrally implantable neurostimulation of choice for chronic pain since the 1970–1980s, with some neurosurgeons favouring motor cortex stimulation as the 'last chance saloon'. This article reviews the available data and professional opinion of the current state of DBS as a treatment for chronic pain, placing it in the context of other neuromodulation therapies. We suggest DBS, with its newer target, namely anterior cingulate cortex (ACC), should not be blacklisted on the basis of a lack of good quality study data, which often fails to capture the merits of the treatment.

Keywords: pain; DBS; ACC

---

1. Introduction

Chronic pain is an important health issue drastically altering the lives of those it affects; it is estimated that 5% of the population suffer chronic pain despite pharmacotherapy [1]. The ramifications include the mental health status of the individual in terms of emotional well-being [2], opioid dependency [3], and cognitive function [4]. The socioeconomic sequelae include loss of productive workforce [5]. In the United States it is estimated to cost $500 billion a year in medical treatment and loss of productivity, with an estimated 116 million people suffering from the condition [6].

Neuropathic pain is defined as "pain caused by a lesion or disease of the somatosensory system" [7]. Chronic pain is that extending beyond the time of injury and healing. Much has been made of the types of pain amenable to different neuromodulation methods. It has been tempting to categorize chronic pain by its cause, and then into categories such as 'nociceptive vs. deafferentation' or 'peripheral vs. central'. The utility of this is questionable, given that the development and maintenance of pain is now thought to involve neuronal plasticity encompassing centrally mediated changes, as suggested by both functional neuroimaging and electrophysiological data [8–13]. The efficacious results of deep brain stimulation (DBS) in spinal cord-related patients, for example, those with failed back surgery syndrome (FBSS), are consistent with this theory, suggesting a centrally mediated component to this initially peripheral injury, which is able to respond favourably to thalamic or anterior cingulate cortex (ACC) stimulation [14]. The lesson here may be to worry less about the specific pain aetiology or the categorization of physical pain, and instead select patients whose pain is not complicated by psychogenic factors such as catastrophization or other negative predictors of good outcome [15].

2. Management of Chronic Pain

2.1. Pharmacotherapy and Non-Invasive Neuromodulation

Pharmacotherapy, such as opioids, carbamazepine, gabapentin, tricyclic antidepressants, and serotonin- and norepinephrine-selective reuptake inhibitors, often fail those afflicted [16]. They cause side-effects such as sedation and nausea because of the non-specificity of the medication, and opioids in particular suffer from reduced long-term efficacy due to receptor downregulation. Others may focus on reducing aberrant neuronal activity in peripheral nerves, failing to address the central nervous system (CNS) aspect involved in its development and maintenance. Neurosurgical attempts to relieve chronic pain focus on the various structures in the pain pathway (peripheral nerve, dorsal root, spinal cord, midbrain, thalamus, and cingulate cortex), either lesioning, electrically stimulating, or perfusing with analgesia/anaesthetic. The opiate epidemic in the USA has forged a renewed interest in neuromodulation. Between 2000–2012, the prescription use of opioids tripled [17]. In 2016, 42,000 Americans lost their lives to opioid overdose, with fentanyl the biggest culprit [18]. Whilst heroin takes second place, it is thought that the indiscriminate prescription of opioids encourages those predisposed to develop addictions, which leads to more recreational drug use [17–19].

Public consciousness is more at ease with less invasive approaches. Non-invasive neuromodulatory strategies include repetitive transcranial magnetic stimulation (rTMS) and transcranial direct current stimulation (tDCS). Both are thought to alter the maladaptive plasticity within pain circuits, affecting the nuclei in the thalamus and subthalamic regions [20–22]. The effects of rTMS are thought to be modest and short-lasting [23]. A systematic review of six studies assessing 127 patients treated with rTMS following spinal cord injury (SCI) concluded that, despite some reduction in the pain indices following rTMS, the effects were unable to reach statistical significance [24]. Of course, the field of non-invasive modulation has its own discrepancies that may cloud the literature; the location (motor cortex versus premotor cortex/dorsolateral prefrontal cortex), type, and orientation of the coil, schedule of repetitive stimulation, and persistence of therapeutic response. rTMS has also been shown to predict the beneficial effects of a more invasive longer-term treatment [25,26], namely motor cortex stimulation (MCS), discussed below, allowing for an exciting area of future research regarding the pre-selection of patients. The second non-invasive approach, tDCS, in contrast to rTMS, does not result in neuronal firing, but changes the resting membrane potential, thereby altering the neuronal excitability. It is thought to alter neurotransmitter systems, hence its longer-term potential [27]. A positron emission tomography (PET) study of 16 SCI pain patients who were administered stimulation over the motor cortex demonstrated not only a reduction in the pain visual analogue scale (VAS) scores, but were found to exhibit an altered metabolism in the subgenual ACC, left dorsolateral prefrontal cortex, and insula, suggesting an effect of tDCS on the affective component of pain processing [28]. As a result of these studies, tDCS is listed as a third-line therapy for neuropathic pain for SCI in the CanPain guidelines [29]; the only neuromodulatory strategy to be included.

2.2. Spinal Cord Stimulation

Spinal cord stimulation (SCS) is a successful and common strategy for treating chronic pain, first reported half a century ago [30,31]. Classically, it is thought to activate the large rapidly conducting Aβ fibres; leading to the potentiation of inhibitory neurons on pain, as per Wall and Melzack's theory [32]. The efficacy of alternative stimulation waveforms such as high frequency and burst, however, suggest that a revision of this theory is needed [33]. For conventional SCS, electrodes are placed with the stimulating tips between the C5 and T1 vertebral bodies for upper limb pain, and between T9 and T11 for lower extremity pain. Traditional 'tonic' stimulation induces a paraesthesia covering the anatomical distribution of the pain, although newer waveforms are paraesthesia-free. SCS is excellent for aetiologies such as FBSS, multiple sclerosis, and complex regional pain syndrome (CRPS), but less effective for phantom limb pain and postherpetic neuralgia [34]. The large number of patients, low morbidities, and improvements in technology over the last 50 years have led to its widespread use. It is the most demonstrably successful neurostimulation method used for chronic pain, largely due to the upsurge of patients with FBSS, present in 10–40% of patients after lumbar spine surgery [35,36].

The first randomised control trial (RCT) looking at SCS for chronic pain used FBSS patients, with the control group receiving repeat lumbar spine surgery. North et al. found a significantly greater number of patients with a 50% or greater pain relief (9/19 for SCS, 3/26 for controls; $p < 0.01$) [37]. The positive results were followed by the PROCESS trial; an RCT of 100 patients; controls received conservative medical management (CMM). SCS delivered better pain outcomes at 6, 12, and 24 months follow up, with the percentage of patients reaching the target of 50% reduction at 24 months being significantly higher; 37% vs. 2% ($p = 0.003$) [38,39]. Indeed, the preliminary findings of a further RCT ($n = 218$) comparing optimal medical management and SCS showed SCS to be superior in the number of patients to reach a 50% pain reduction [40].

Success with other pain aetiologies has been published, but is less convincing. Kemler et al. demonstrated that patients with complex regional pain syndrome (CRPS) experienced a mean VAS reduction of 2.4/10 cm ($n = 24$) at a six month follow up, and 3.6/10 cm ($n = 18$) for those with physical therapy plus SCS compared to an increase in 0.2 cm/10 cm at six months for physical therapy alone [41]. These values, however, do not include the 12 patients from the SCS arm who failed to complete the implant procedure because of an unsuccessful test stimulation. At a five-year follow-up, the pain relief differences did not reach significance. [42]. Two prospective studies suggest that SCS improves painful peripheral neuropathy compared to medical management with 60% of patients in the SCS group reaching the success criteria at six months, compared to 5% and 7% in the control arm [43,44].

The level of efficacy for SCS in chronic pain is still classed as 'moderate', but importantly, it is shown to be safe with a 2005 systematic review of studies showing no major adverse events [45]. Novel devices are demonstrating further potential. High frequency versions of the treatment provide up to 10,000 Hz (compared to up to 1200 Hz) and have demonstrated positive results in feasibility studies [46]. These versions have even been shown to be more successful compared to the conventional frequencies in a study of 193 subjects (171 of whom completed implantation) of back pain (84.5% at three months with high frequency compared to 43.8% with conventional) and leg pain (83.1% vs. 55.5%; $p < 0.001$) [17,48]. Burst DR® stimulation (Abbott, One St Jude Medical Drive, St Paul, MN 55117, USA) has also enjoyed additional success with hints it might improve on conventional SCS outcomes with similar safety profiles in both FBSS [49] and CRPS [50]. Dorsal root ganglion (DRG) stimulation has some theoretical advantages over conventional SCS; it delivers stimulation directly to the nerve roots and is less vulnerable to positional changes [51], as such initial studies have shown positive results in the regions not usually successful with SCS, for example CRPS and groin pain. The prospective RCT that led to United States Food and Drug Administration approval of its use demonstrated that at a three month follow up 81.2% of CRPS/causalgia patients treated with DRG achieved success (defined by greater than 50% reduction in VAS score) with conventional stimulation success at 51.7% [52]. However, higher procedural adverse events were found in the DRG group, such as pain at the incision site (7.9% DRG and 6.9% SCS).

Patient selection is key, as for all of the neurostimulation procedures, with risk factors for negative results, including opiate addiction, catastrophization, active depression/anxiety, low-activity levels, as well as ongoing litigation, necessitating a role for presurgical neuropsychological evaluation [53]. Currently, rTMS promises insight into finding the suitable candidates [54]. If SCS fails, or if the pain aetiology is central (e.g., post-stroke pain and atypical facial pain), a surgeon may try either DBS or MCS, largely depending on their skillset and familiarity, as has been shown to be the case in several studies [55–57].

2.3. Deep Brain Stimulation

Deep brain stimulation provides a further opportunity to alleviate pain in some individuals. Specific indications include central post-stroke pain, atypical facial pain, brachial plexus injury, and some patients who have failed SCS. Such conditions do not generally respond to SCS or DRG stimulation, except for some cases of facial pain that may respond to peripheral nerve stimulation or high cervical stimulation [58,59]. There are three main DBS target sites, namely: (1) the thalamus-ventral posterolateral nucleus and ventral posteromedial nucleus (VPL/VPM); (2) regions surrounding the third ventricle and aqueduct of Sylvius, including the grey matter- Periventricular grey and periaqueductal grey (PVG/PAG); and (3) the newer target of the rostral anterior cingulate cortex (ACC) posterior to the anterior horns of the lateral ventricles. A fourth target, the posterior hypothalamus, may be considered specifically for cluster headache (not discussed in detail in this article), in those for whom occipital nerve stimulation has failed. For the thalamus and PAG, the DBS at lower frequencies (<50 Hz) is thought to be analgesic and at higher frequencies (>70 Hz) is thought to be hyperalgaesic. When VPL/VPM is targeted, pleasant paraesthesia supplants a painful sensation, whereas PVG/PAG stimulation induces a sense of warmth and analgesia over the area of pain [60–62]. ACC stimulation is thought to remove the affective aspect of pain, and thus high frequencies have shown to be clinically effective [63]. The exact mechanism of action is equivocal [64]. There are mixed reviews regarding endorphin and opioid pathway theories, with earlier studies showing these to be less likely [65,66]. However, more recent positron emission tomography (PET) studies demonstrate a reduced binding of (^{11}C)diprenorphine (a ligand with high opioid affinity) in the dorsolateral PAG when DBS electrodes were switched on vs. off, indicating the DBS-induced PAG release of endogenous opioid peptides [67–69]. However, Pereira et al. have shown elevated gamma band frequency in PAG/PVG upon the administration of naloxone in DBS patients [70], suggesting an enhanced awareness of the patient's worsening pain. Other studies suggest that both DBS and SCS may modulate the gene expression [71,72].

A brief history of the evolution of DBS for chronic pain helps to contextualise how such an invasive procedure becomes relevant to the field of chronic pain. The ability of DBS to alleviate pain was first seen in septal self-stimulation experiments in rodents [73]. The first DBS was performed for nociceptive pain in the 1950s [74]. The 1960s saw use of DBS to alleviate pain in cancer patients [75]. Further impetus for electrical stimulation, initially in the form of peripheral nerve stimulation [76] and then spinal cord [30] stimulation, was found in Melzack and Wall's gate theory of pain [32]. By the 1970s, several centres were performing DBS for neuropathic pain. Evidence for targeting PVG/PAG came from pain relief in rodents during awake surgery [77,78], and human studies followed [79–82]. Evidence for targeting the thalamic nuclei (VPL/VPM) came from ablative surgery [83–85], with subsequent human studies [86–90], along with Adams, who also targeted the internal capsule [91–93], moving to more medial targets developed from localisation errors and investigations in the current spread [94–97]. By 1987, out of the 141 patients implanted, 59% obtained initial pain relief, although this percentage reduced to 31% at follow up (mean 80 months) [98]. The major complications were listed as one death, 12% wound infections, and 3.5% intracranial haemorrhage.

The initial human studies lacked numbers and were marred by the variability of the surgical technique or settings, different locations used, and different pain profiles of the patients being treated, leading to heterogenous patient groups, and hence the studies were not successful. An early RCT by Marchland showed placebo improved pain intensity, yet the stimulation of the thalamus did not [99]. Two failed industry open label studies further dampened the excitement. The first (n = 196), by Medtronic, was powered to show if half of the patients that were internalised would get at least 50% pain relief. This failed to reach the outcome. The second (n = 50) trial failed because of the lack of accrual [100]. Consequently, the FDA afforded only 'off-label' status to DBS for chronic pain relief [101]. Few clinical trials have been reported since, despite the apparent need to rectify the multitude of issues plaguing the above trials. These issues include the nonrandomised nature, heterogenous case mix, subjective and unblinded assessment of patient outcome, inconsistencies in DBS sites stimulated, stimulation parameters chosen, and number of electrodes implanted.

Despite this relegation to the 'off-label' status, the DBS for chronic pain has yielded many success stories, arguably not all represented in the literature. Boccard et al. reported the long-term outcome of 59 patients with a variety of pain aetiologies receiving DBS in the sensory thalamus, periventricular grey, or both. After a mean follow-up of almost 20 months, the pain was compared to the pre-operative levels. Improvement was defined as a global improvement of their EuroQol-5D (EQ-5D). For patients with phantom limb, 8/9 improved; for brachial plexus injury, 3/6 improved; for post-stroke pain, 16/23 improved; for spinal cord injury, 4/7 improved; and for cephalalgia, 6/11 improved [102]. For the patients that improved, the pain was reduced by 50% on a visual analogue scale.

In another study of mixed pain aetiology, Kumar et al. reported the outcome of DBS in the periventricular/periaqueductal grey area ($n = 49$) or sensory thalamus/internal capsule ($n = 16$) [103]. Mean follow-up was 78 months and success defined as a greater than 50% reduction in VAS pain scores. For the patients with FBSS, 32/43 had long-term improvement; for peripheral neuropathy, 3/5 improved; for thalamic pain, 1/5 improved; for trigeminal neuropathy, 4/4 improved; for spinal cord injury, 0/3; for post-herpetic neuralgia, 0/3; and for phantom limb pain, 1/1 improved.

A 2005 meta-analysis listed differing success rates depending on aetiology. Most of the studies included defined pain in the normative way, of "at least 50% of patients with a 50% improvement in pain scores". Some pain aetiologies fared better than others. For example, FBSS has a 78% success rate (pre-internalisation), causalgia 80%, cancer 65.2%, lumbosacral radiculopathy 90.5%, and lumbar arachnoiditis 77.8%; however thalamic central lesion, phantom limb, and cervical root/brachial plexus lesion scored less well (31.1%, 44.4%, and 50%, respectively). The authors calculated the total success rate overall to be 232/424 (54%) of the surgeries, of which the successful ones were internalised. In the post-internalisation, 76.1% remained successful [61]. Rasche et al. further demonstrate the importance of careful patient selection, agreeing that DBS appears to be particularly useful for FBSS, although it shows disappointing results for SCI and poststroke pain [104].

In the past decade, to our knowledge, there have been only three clinical trials pertaining to DBS and pain. Two are prospective randomised crossover trials and one [105] is a non-randomized open label trial that serves as a comparison between MCS and DBS (discussed below). The more tangential of the RCTs targeted the subthalamic nucleus in 19 Parkinson disease patients to show that the post-operative pain threshold increased with no correlation between the increased pain threshold and improvements in the UPDRS-3 scores, thus suggesting that clinical pain alleviation after subthalamic nucleus-DBS is not just a by-product of the improvement in motor complications in these Parkinson patients [106].

The third RCT in a 2017 trial using post-stroke pain patients targeted the ventral striatum/anterior limb of the internal capsule. Ten patients, nine of whom progressed to internalisation, were implanted, waited one month, and then randomised to either 'active' or 'sham' stimulation for three months, after which they crossed over to the other treatment category. The results show no significant difference in pain-related-disability as indexed by the arbitrarily set 'greater than 50%' improvement on the pain disability index (11% DBS on vs. 12% DBS off; odds ratio = 1.05, 95% CI 0.96–1.15 $p = 0.270$). However, the authors highlight an acceptable safety profile, with 14 serious adverse events recorded and resolved, only three of which (one seizure, one wound dehiscence, and one wound infection) were identified as being related to study. They also found statistically significant improvements on multiple outcome measures related to the affective sphere of pain, for example, a 50% improvement on MADRAS (Montgomery–Asberg Depression Rating Scale); with 44% reaching the 50% reduction target with DBS on vs. 19% with DBS off; odds ratio 0.30, 95% CI 0.11–0.83; $p = 0.020$. No significant difference was seen for the VAS reductions between the on and off states [107].

These modest results may reflect the inability of the data to represent the potential for DBS for a number of reasons. DBS tends to be a treatment that takes place once SCS has failed, suggesting that the patient population that received DBS are filtered to be those that are more difficult to treat than those received by SCS, skewing the results unfavourably against DBS. It is also pertinent that for some of those patients who do not meet the arbitrary '50% reduction in pain' threshold, testimonials suggest that even a partial reduction in pain has resulted in a greatly increased quality of life. Put simply, there are patients in whom DBS has reduced their pain and improved their subjective quality of life, but who are represented as a 'fail' in the literature. Reductions in VAS are poorly correlated with patient satisfaction or disability [108]. In fact, in this study, 5/9 patients said they would have the surgery all over again if they knew the result they would get—suggesting over 50% success rate according to patient satisfaction. There are further issues with the outcome measures used to detect successful results; the removal of a particular component of pain, for example, burning hyperesthesia, may unmask another type, such as muscular allodynia, as has been described after stroke [109]. In the future, a score capturing a more objective measure of the changes in analgesia may be useful. Investigations into the heart rate variability changes and blood pressure monitoring may provide an objective measure of efficacy that correlates to analgesia [110,111].

Furthermore, the levels of success rates would be easily increased following improvements in patient selection, in predicting who will respond. Diffusion tensor imaging (DTI) techniques (i.e., looking at network connectivity to predict a response) may provide a solution, as has been suggested from work with movement disorder patients. In these patients, it is thought that predominant beta-activity may come to serve as an electrophysiologically determined target for the optimal outcome in the subthalamic nucleus for Parkinson's disease [112].

Fundamentally, RCTs examine population statistics—they look at mean changes. For an individual with refractory pain who has used up the limited treatment options available, it is a question of risk vs. benefit, for example, an individual may be weighing up a 20% chance of success with a 1:500 risk of stroke. It is reasonable the individual may choose to take this risk.

3. Anterior Cingulate Cortex: A More Recent DBS Target for Chronic Pain

Pain relief by cingulotomy [113–115] ignited interest in the dorsal ACC as a potential DBS target in the treatment of chronic refractory pain, especially for those with a substantial affective component to pain. Foltz and White built on the observations of psychiatric lobotomy patients to suggest that the transection of the cingulum bundle might benefit those patients with a disproportionally large affective component to their pain [116]. In the discussion of their findings, they reported a universal decrease in the distress associated with chronic pain. Ballantine and Hurt later introduced a stereotactic approach in 1966 [117,118]. This target is supported by more recent imaging and neurophysiological evidence describing the role of the anterior cingulate in the perception of pain [119]. Studies using functional Magnetic Resonance Imaging (fMRI) have demonstrated an increased activation of the dorsal anterior cingulate cortex (dACC) during both empathic and experienced pain [120], supporting the notion that the dACC is implicated in the affective component of pain. Furthermore, PET studies of thalamic DBS patients ($n = 5$) demonstrate the activation of anterior ACC throughout the 40 min of DBS tested, and a more posterior ACC activation at a delay (approximately 30 min) after onset [121].

In 2007, Spooner and colleagues, reported the first case in which standard DBS electrodes were used to administer high-frequency electrical stimulation of the dACC. The patient had refractory neuropathic pain resulting from a complete C4 level spinal cord injury, despite numerous medical and surgical interventions. Targets were place in the cingulate cortex (bilaterally) and PVG (unilateral), and a one-week blinded stimulation period was completed. The outcome measures included VAS, pain medication usage, and self-described mood. Both the PVG and ACC stimulation decreased the VAS pain rating and led to a reduction in the subcutaneous lidocaine usage [122]. At the three-month follow up, the ACC stimulation yielded a VAS score of three (out of 10) and a mood described as 'best'. In comparison, PVG scored a value of four with an 'average' mood, and no stimulation resulted in a VAS score of 10 and a mood self-described as 'worst'. The improvement in mood and reduced pain with bilateral cingulate stimulation implies that the affective component of pain is targeted. A series of cases from the Oxford group have since replicated these benefits in case series ranging from n-of-1 to n-of-24, demonstrating that bilateral ACC stimulation is not only efficacious for a variety of pain aetiologies including FBSS, post-stroke, spinal cord injury, brachial plexus lesions, and head injury, but that it delivers long term control over a period of years [63,123,124]. In the most recent case-series, 83% of patients showed an improvement in numerical rating scale (NRS) pain score by at least 60% ($p < 0.001$) at the six month follow up, and the McGill pain questionnaire (MPQ) decreased by 47% ($p < 0.01$). After one year, the NRS score decreased by 43% ($p < 0.01$), the EQ-5D quality of life measure was significantly reduced (mean, -30.8; $p = 0.05$), and significant improvements were also observed for different domains of the short form health survey (SF-36). At the longer follow-up, the efficacy was sustained up to 42 months in some patients, with an NRS score as low as three [63]. Importantly, the patients described that although pain was present, it was 'less bothersome' or 'separate from them', playing into the affective role of the ACC. This suggests that the NRS scores are not necessarily capturing the patient satisfaction of the treatment. Of note, four patients experienced problems with seizures/epilepsy after long-term stimulation, one of whom suffered from breakthrough seizures despite being off-stimulation and taking anti-epileptics. The ethical dilemma this poses has been discussed in a recent publication [125]. If this risk can be minimized, it is possible that ACC may be able to salvage patients in whom other neuromodulation has failed. However, given the small number of patients and short follow up time compared to other DBS targets, we must be cautious regarding the projections for its future use. The case studies and series pertaining to DBS targeting ACC for chronic pain are summarised in Table 1.

Table 1. Case studies and series of Deep Brain Stimulation targeting Anterior Cingulate Cortex.

Paper	Article Type	Patient N	Aetiology of Pain	Target	Outcome Measures	Follow up Times	Results	Conclusion
Boccard, Prangnell et al. (2017) [63]	case series	24	[a]FBSS (6), post-stroke (9), SCI (2), brachial plexus injury (3), unknown chest pain (1), head injury (1), [b]RTA (2).	[c]ACC (bilateral)	[d]NRS, [e]SF-36, EQ-[f]5D, [g]MPQ	6 months, 1 year, 12 people f/u at 38.9, some at 42 months	At 6 months. NRS decreased from 8 to 4.27 ($p = 0.004$), MPQ improved (mean -36%; $p = 0.021$), EQ-5D score decreased (mean -21%; $p = 0.036$). The physical functioning domain of SF-36 was significantly improved (mean $+54.2\%$; $p = 0.01$). At 1 year NRS score decreased by 43% ($p < 0.01$). EQ-5D reduced (mean -30.8; $p = 0.01$). Improvements in domains of SF-36. At the longer f/u; efficacy was sustained up to 42 months. NRS score as low as 3.	ACC stimulation alleviates chronic neuropathic pain refractory to pharmacotherapy.
Boccard, Fitzgerald et al. (2014) [124]	case series	16 (15 internalized; 11 followed up)	FBSS (6), Post-stroke (4), Spinal Cord Injury (1), Brachial plexus (3), unknown chest (1), head injury (1)	ACC (bilateral)	[h]VAS, SF-36, EQ-5D, McGill Pain Questionnaire	mean 13.2 months	Post-surgery, VAS decreased to <4 in five patients, and one patient reported to be pain free. Significant improvements on EQ-5D observed (mean 20.3%; range 0%–83%; $p = 0.008$). Statistically significant improvements were observed for the physical functioning and bodily pain domains of SF-36 quality of life survey; mean $+64.7\%$ (range, -8.9%–$+27\%$; $p = 0.015$) and mean $+39.0\%$ (range -33.8%–$+159\%$; $p = 0.05$).	ACC DBS can relieve chronic neuropathic pain refractory to pharmacotherapy and restore quality of life.
Boccard, Pereira et al. (2014) [123]	case study	1	RTA/brachial plexus injury	ACC (bilateral implants)	VAS, SF-36, McGill pain questionnaire, EQ-5D, Neuropsychological measures	2 years post-surgery	VAS decreased from 6.7 to 3; McGill improved by 43%, EQ-5D Health state increased by 150%.	ACC DBS efficacious; ACC target has potential for long-term control
Spooner, Yu et al. (2007) [122]	case report	1	Spinal Cord Injury at C4	ACC (bilateral); [i]PVG (unilateral)	VAS, pain medication usage, described mood.	1–5 days post-surgery, 4 months post-surgery, 1 year not possible (patient died due to pulmonary issues)	Results most striking at 3 months with cingulum stimulus scoring VAS 3 and lidocaine usage of 2 (cc/hr), mood described as 'best'. Compared to PVG (VAS 4, lidocaine 2, mood 'average') or no stimulation (VAS 10, lidocaine 5, mood 'worst').	Bilateral cingulate stimulation improved the patient's mood and reduced pain more completely than PVG stimulation or medication alone

[a]FBSS = Failed Back Surgery Syndrome. [b]RTA = Road Traffic Accident. [c]ACC = Anterior Cingulate Cortex. [d]NRS = Numeric Rating Scale. [e]SF-36 = Short-form 36 quality of life questionnaire. [f]EQ-5D = EuroQol 5-Domain quality of life questionnaire. [g]MPQ = McGill Pain Questionnaire. [h]VAS = Visual Analogue Scale. [i]PVG = Periventricular Grey.

4. Motor Cortex Stimulation and Its Comparison with Deep Brain Stimulation

There has been much discourse surrounding the need to compare the less invasive motor cortex stimulation (MCS) with DBS in terms of efficacy and the side-effect profile. This discourse is yet to filter down to good quality studies of clinical relevance.

To briefly describe MCS, epidural electrodes are implanted over the motor cortex through a frontoparietal craniotomy. One or two electrodes are implanted, either parallel or orthogonal to the central sulcus, to comply with the motor representation of the painful area. Then, similar to DBS, the electrode is connected to a subcutaneous implant pulse generator, with stimulation parameters adjusted post-operatively.

The MCS studies show varying levels of success. In a summary of the literature surrounding MCS for chronic pain up to 2006, a greater than 40% improvement in pain scores were reported in 54% of 117 patients with central pain, and 68% of 44 patients with trigeminal neuropathic pain [101]. A prospective audit of 10 patients with mixed pain aetiologies showed a 50% success rate (relief of pain between 50–90% from baseline), with no clear predictability based on the mixed pain aetiologies [126]. Later results have shown mixed success. Two studies have effectively reported MCS to be ineffective, but are hampered by issues surrounded by patient retention, patient selection, and administration of the treatment [127,128]. Lefaucher et al. reported an RCT of MCS for peripheral neuropathic pain, where 13 patients had a significant reduction in some measurements of pain when the device was 'on' compared to 'off'. However, these results were statistically insignificant after multiple comparison correction [129]. Nguyen et al. reported a randomized, blinded crossover trial of MCS in 10 patients with neuropathic pain with significant reduction in pain when the device was switched 'on' compared to 'off' [130]. Notably, a disappointing response was seen in hemibody post-stroke pain and post-herpetic neuralgia patients. The poor results for the post-herpetic neuralgia have been replicated [131]. The same group reported better results for patients with complex regional pain syndrome, however, with four out of five patients experiencing improvement with pain, sensory, and sympathetic symptoms [132]. Results concerning trigeminal neuropathic pain also appear more successful, with Rasche et al. demonstrating that 5 out of 10 patients received a reduction in VAS pain scores of at least 50 [133]. By 2012, a review of the MCS facial pain literature showed that an impressive 84% of 100 patients implanted following a trial had at least 40% pain improvement [134]. This success continues to be replicated, with 72% of 36 patients receiving MCS for trigeminal neuropathic pain showing a mean VAS reduction from 8.11 to 4.5 cm ($p < 0.05$) and a mean VAS score of 5 cm at the last follow up (mean 5.6 years and 26 patient included) [135].

Table 2 lists the studies, to our knowledge, involving a MCS vs. DBS comparison. As shown, the studies are sparse, and with a superficial glance, may seem to favour MCS—this, however, is far from clear cut. The study by Son et al. appears to favour MCS. Nine chronic pain patients of varying aetiologies were implanted with both MCS and thalamic DBS; 6/8 responded better to MCS and 2/8 responded to DBS [105]. They concluded MCS to be a reasonable initial means of treatment given the less invasive nature and the lack of evidence showing DBS to be of higher efficacy. This conclusion is perhaps premature, as the majority of DBS for chronic pain no longer involves solely a thalamic implant. It would be prudent to conduct similar studies with DBS implants in the PAG in addition to thalamus, or alternatively, the ACC if appropriate, in order to avoid a 'MCS gold standard' vs. 'DBS old standard' comparison. Ideally this type of RCT study would be replicated for a wider variety of pain aetiologies, specifically using DBS targets for which there is more experience, such as thalamus/PAG/ACC. In a review of the MCS and DBS literature, Honey et al. suggest that, in addition to having 'pure cohorts' comparing DBS and MCS in one condition at a time, future trials should incorporate a post-operative phase, where the success of each modulation type can be maximized, followed by a blind crossover phase to test response, and finally, an open-label phase to monitor long term efficacy [136].

Table 2. Studies involving comparison of Motor Cortex Stimulation and Deep Brain Stimulation.

Paper	Article Type	Patient N	Aetiology of Pain	Target	Outcome Measures	f/u Times	Results	Conclusion
Nandi et al. (2002) [57]	case-series	10	All post-stroke pain. [a]MCS patients: post-stroke hemi-body pain (4); post-stroke facial pain (4); [b]DBS patients: post-stroke hemi body (3), post-stroke face and leg (1)	[c]PVG	[d]VAS	2–3 weeks; some up to 4 years	MCS: 1/6 success rate. DBS:3/4 had at least 40% reduction in VAS scores during stimulation, 2/2 internalised with success.	MCS is not effective relieving post-stroke neuropathic pain. DBS is the preferred option.
Katayama et al (2001 a.) [55]	case-series	45	phantom limb (trauma- [e]rt leg), brachial plexus avulsion (rt arm).	thalamus	VAS	unspecified- results reported to be 'long term'	All 19 patients were given [f]SCS and if failed were split into either DBS or MCS. For DBS 60% (6/10) gave pain relief, and for MCS 1/5 (20%) required pain relief. 4 patients were given both DBS and MCS- one patient reported better pain control by MCS than DBS. 2 patients reported the opposite.	DBS preferable to MCS, especially lower limb.
Katayama et al (2001 b.) [56]	case-series	45	post-stroke pain	thalamus	VAS	unspecified- results reported to be 'long term'	Success rates (defined as >60% reduction in VAS scores) of 7% for SCS (3/45), 25% for DBS (3/12), 48% for MCS (15/31)	Success rate increases as stimulation moves higher. MCS more successful than DBS.
Son, Kim et al. (2014) [105]	open label	9*	Central post-stroke pain (4), [g]SCI (4), amputation stump pain in arm (1)	ventralis caudalis (Vc) thalamus DBS	[h]NRS, medication use.	39 months mean, (8–72)	6/8 (75%) responded to MCS. 2/8 had successful DBS (one patient with amputation stump pain and the other with SCI pain caused by cervical syrinx). NRS score decreased significantly ($p < 0.05$) MCS: 37.9 ± 16.5 and DBS 37.5%.	Considering the initial success rate and the less invasive nature of epidural MCS compared with DBS, MCS would be a more reasonable initial means of treatment for chronic intractable neuropathic pain.

[a]MCS = Motor Cortex Stimulation. [b]DBS = Deep Brain Stimulation. [c]PVG = Periventricular Grey. [d]VAS = Visual Analogue Scale. [e]rt=right. [f]SCS = Spinal Cord Stimulation. [g]SCI = Spinal Cord Injury. [h]NRS = Numeric Rating Scale. * = 8 successfully implanted and used in the comparison.

For any given study, it is reasonable to suppose that the research group and/or surgical team in question is better skilled at one type of intervention over the other, producing unintentionally biased data. This may go some way to explaining why, even when controlling for pain type (i.e., just focussing on post-stroke pain patients), Katayama found MCS to yield higher success rates than DBS [56], whereas the opposite was seen in a study from a different group [57], where three out of four patients showed significant difference in VAS scores, compared to 3/6 MCS patients who experienced no pain relief. It is also possible that the relative merits of DBS and MCS may change depending on the specific aetiologies of the pain. For example, Katayama showed that for those with post-stroke pain, a patient may wish to opt for MCS over DBS; 48% (15/31) vs. 25% (3/10) success rate [56], but regarding phantom limb patients, DBS prevails with a success rate of 60% (6/10) in the DBS, but 20% (1/5) in MCS [55].

Safety comparisons between the two methods are much needed. The data is lacking, with the findings either unreported [55,56] or studies being simply too small to make a real comparison. To illustrate with Nandi et al., one DBS patient suffered a CSF leak and haematoma over the generator, whilst the MCS patient morbidity included one subdural haematoma and secondary wound infection, one seizure induced during post-op titration, one with a 'strong motor response' elicited during the procedure, and one patient 'affected by the magnetic field' with no further details given [57]. Whilst Nandi lists the adverse events, it is difficult to unpick a sensible conclusion, particularly if surgical skills and experience vary between the different modalities tested.

Explanations of why one patient is fielded into the MCS camp rather than DBS and vice-versa, are not provided. Furthermore, patient groups sometimes differ in baseline characteristics (Nandi et al. showed age characteristics of 59.5 years in the MCS group and 70.5 in the DBS group). Moreover, current publications are comparing MCS to a metaphorical and literal 'moving target' as electrode implants vary in locations. The targets of DBS may yield different results, such that sometimes the comparison is MCS vs. DBS:thalmus, other times MCS vs DBS: PAG/PVG and soon studies may compare MCS vs. DBS:ACC. This is particularly relevant if, as one meta-analysis demonstrates, the outcomes are more successful in those patients with targets in thalamus and PAG together, as compared to those patients with targets in the thalamus only; 87.3% success rate vs. 58%, respectively ($p < 0.05$) [61].

5. Conclusions

The use of neuromodulation for chronic pain has helped many patients, for whom pharmacotherapy has failed. SCS has been shown to be particularly useful for those with FBSS, but its success extends beyond this. DBS appears to fall short upon a review of the literature, possibly a misrepresentation of the innate potential of DBS as a treatment for chronic pain. The aforementioned methodological limitations of the published studies, together with the difficulty of comparing the efficacy to its closest alternatives, clouds the potential of DBS. On aggregate, it is difficult to draw conclusions from non-randomised trials that are understandably limited by small sample size. It may be that good-quality RCTs are not realistically achievable because of the cost of treatment and the rarity of its use. One alternative here is to use Bayesian statistics as has been suggested for rare disease groups. This approach provides probabilities of treatment effects of various percentages that can be applied to the next patient similar in clinical problems [137].

It is true that DBS does not reduce pain in all patients, and sometimes produces unwanted, mostly manageable, side effects. It is also true that many patients treated with DBS for chronic pain have been satisfied with their pain reduction, even some of those classed as a 'failed treatment' in the literature. Indeed, the National Institute of Clinical Excellence (NICE) guidelines do appreciate this, approving DBS for chronic refractory pain where other methods have failed and a multidisciplinary team of pain specialists approve of the case (IPG382). Although DBS for chronic pain is not currently funded on the National Health Service in the United Kingdom, it would seem that the world is experiencing a renaissance of the exploration of DBS for chronic pain; there are currently ten clinical trials registered regarding pain and DBS, with the United States claiming five of these, France four, and Denmark one suspended study (clinicaltrials.gov; nil found on EudraCT).

The results from previous papers and reviews suggest that the DBS for chronic pain is most successful for pain after amputation (both phantom limb and stump), FBBS, cranial and facial pain including anaesthesia dolorosa, and plexopathies. Poststroke pain is particularly successful if the type of pain reported is burning hyperaesthesia [138–140]. Given that we are dealing with refractory pain, rather than trying to 'prove' or 'disprove' the efficacy of DBS in large patient populations, perhaps it is more appropriate to adopt a treatment pathway that first uses less invasive therapies, followed by SCS or peripheral stimulation (if appropriate), and then DBS or MCS. Current pain theories would suggest over time, all of the pain circuits become centrally mediated, suggesting that a one-size-fits-all view may not be as inappropriate as previously thought, shifting the emphasis away from particular pain aetiologies being either amenable or unresponsive to DBS. It remains to be seen how the introduction of ACC as a target will change success rates for chronic pain patients, and how this will alter comparisons of DBS and MCS, as we await longer-term follow up data and increasing patient numbers. The targeting of an affective process promises a catch-all for those pain aetiologies that have proved more troublesome for less-invasive techniques. The obvious disadvantage being a possible risk of seizures/epilepsy.

The ability to pre-select individuals who respond well to a particular neuromodulation would lead to better outcomes. We are currently far from this patient-specific pre-selection ability, but some tantalising hints have proved simultaneously exciting and frustrating. Evidence from LFP recording shows chronic pain patients with DBS 'off' have characteristically enhanced low frequency (8–14 Hz) power spectra of both PAG and VP (thalamus) local field potentials when in pain [141]. Further research could explore non-invasive functional neuroimaging, including single-photo emission computed tomography, PET, and MEG to find correlates of this [109,142–144]. Perhaps rTMS may be an aid to selection as it can be with SCS and MCS. Meanwhile, there are suggestions that optimizing stimulation parameters post-surgery "through recursive testing and adjustments" leads to pain-improvement, with some evidence demonstrating optimal relief for two test patients with midbrain electrodes whilst cycling 2 Hz on for 1 s and off for 2 s, discovered during comprehensive meticulous parameter testing [145]. The possibilities for improving patient selection and success rates make this an exciting field of both research and clinical practice.

Author Contributions: S.M.F. performed original draft preparation. A.G., T.A., and S.M.F. performed writing (reviewing and editing).

Funding: Research for this review article received no external funding.

Acknowledgments: The research was supported by the National Institute for Health Research (NIHR) Oxford Biomedical Research Centre based at Oxford University Hospitals NHS Trust and University of Oxford.

Conflicts of Interest: The authors declare no conflict of interest.

References

1. Bouhassira, D.; Lantéri-Minet, M.; Attal, N.; Laurent, B.; Touboul, C. Prevalence of chronic pain with neuropathic characteristics in the general population. *Pain* **2008**, *136*, 380–387. [CrossRef] [PubMed]
2. Elliott, A.M.; Smith, B.H.; Penny, K.I.; Smith, W.C.; Chambers, W.A. The epidemiology of chronic pain in the community. *Lancet* **1999**, *354*, 1248–1252. [CrossRef]
3. Gureje, O.; Von Korff, M.; Simon, G.E.; Gater, R. Persistent pain and well-being: A world health organization study in primary care. *JAMA* **1998**, *280*, 147–151. [CrossRef] [PubMed]
4. McCracken, L.M.; Iverson, G.L. Predicting complaints of impaired cognitive functioning in patients with chronic pain. *J. Pain Symptom Manag.* **2001**, *21*, 392–396. [CrossRef]
5. Stewart, W.F.; Ricci, J.A.; Chee, E.; Morganstein, D.; Lipton, R. Lost productive time and cost due to common pain conditions in the us workforce. *JAMA* **2003**, *290*, 2443–2454. [CrossRef] [PubMed]
6. Pizzo, P.A.; Clark, N.M. *Relieving Pain in America: A Blueprint for Transforming Prevention, Care, Education, and Research*; National Academies Press: Washington, DC, USA, 2011.

7. Jensen, T.S.; Baron, R.; Haanpää, M.; Kalso, E.; Loeser, J.D.; Rice, A.S.; Treede, R.D. A new definition of neuropathic pain. *Pain* **2011**, *152*, 2204–2205. [CrossRef] [PubMed]
8. Melzack, R.; Coderre, T.J.; Katz, J.; Vaccarino, A.L. Central neuroplasticity and pathological pain. *Ann. N. Y. Acad. Sci.* **2001**, *933*, 157–174. [CrossRef] [PubMed]
9. Apkarian, A.V.; Bushnell, M.C.; Treede, R.D.; Zubieta, J.K. Human brain mechanisms of pain perception and regulation in health and disease. *Eur. J. Pain* **2005**, *9*, 463–484. [CrossRef] [PubMed]
10. Anderson, W.S.; O'Hara, S.; Lawson, H.C.; Treede, R.D.; Lenz, F.A. Plasticity of pain-related neuronal activity in the human thalamus. *Prog. Brain Res.* **2006**, *157*, 353–364. [PubMed]
11. Coderre, T.J.; Katz, J.; Vaccarino, A.L.; Melzack, R. Contribution of central neuroplasticity to pathological pain: Review of clinical and experimental evidence. *Pain* **1993**, *52*, 259–285. [CrossRef]
12. Schweinhardt, P.; Lee, M.; Tracey, I. Imaging pain in patients: Is it meaningful? *Curr. Opin. Neurol.* **2006**, *19*, 392–400. [CrossRef] [PubMed]
13. Stern, J.; Jeanmonod, D.; Sarnthein, J. Persistent eeg overactivation in the cortical pain matrix of neurogenic pain patients. *Neuroimage* **2006**, *31*, 721–731. [CrossRef] [PubMed]
14. Cruccu, G.; Aziz, T.Z.; Garcia-Larrea, L.; Hansson, P.; Jensen, T.S.; Lefaucheur, J.P.; Simpson, B.A.; Taylor, R.S. Efns guidelines on neurostimulation therapy for neuropathic pain. *Eur. J. Neurol.* **2007**, *14*, 952–970. [CrossRef] [PubMed]
15. Pereira, E.A.; Green, A.L.; Aziz, T.Z. Deep brain stimulation for pain. *Handb. Clin. Neurol.* **2013**, *116*, 277–294. [PubMed]
16. Wilkinson, H.A.; Davidson, K.M.; Davidson, R.I. Bilateral anterior cingulotomy for chronic noncancer pain. *Neurosurgery* **1999**, *45*, 1129–1134. [CrossRef] [PubMed]
17. Jones, C.M.; Logan, J.; Gladden, R.M.; Bohm, M.K. Vital signs: Demographic and substance use trends among heroin users—United States, 2002–2013. *MMWR Morb. Mortal. Wkly. Rep.* **2015**, *64*, 719–725. [PubMed]
18. Dowell, D.; Haegerich, T.M.; Chou, R. Cdc guideline for prescribing opioids for chronic pain—United States, 2016. *JAMA* **2016**, *315*, 1624–1645. [CrossRef] [PubMed]
19. Ranapurwala, S.I.; Naumann, R.B.; Austin, A.E.; Dasgupta, N.; Marshall, S.W. Methodologic limitations of prescription opioid safety research and recommendations for improving the evidence base. *Pharmacoepidemiol. Drug Saf.* **2018**. [CrossRef] [PubMed]
20. Moreno-Duarte, I.; Morse, L.R.; Alam, M.; Bikson, M.; Zafonte, R.; Fregni, F. Targeted therapies using electrical and magnetic neural stimulation for the treatment of chronic pain in spinal cord injury. *Neuroimage* **2014**, *85 Pt 3*, 1003–1013. [CrossRef]
21. Strafella, A.P.; Vanderwerf, Y.; Sadikot, A.F. Transcranial magnetic stimulation of the human motor cortex influences the neuronal activity of subthalamic nucleus. *Eur. J. Neurosci.* **2004**, *20*, 2245–2249. [CrossRef] [PubMed]
22. Lang, N.; Siebner, H.R.; Ward, N.S.; Lee, L.; Nitsche, M.A.; Paulus, W.; Rothwell, J.C.; Lemon, R.N.; Frackowiak, R.S. How does transcranial dc stimulation of the primary motor cortex alter regional neuronal activity in the human brain? *Eur. J. Neurosci.* **2005**, *22*, 495–504. [CrossRef] [PubMed]
23. Papuć, E.; Rejdak, K. The role of neurostimulation in the treatment of neuropathic pain. *Ann. Agric. Environ. Med.* **2013**, *1*, 14–17.
24. Gao, F.; Chu, H.; Li, J.; Yang, M.; DU, L.; Chen, L.; Yang, D.; Zhang, H.; Chan, C. Repetitive transcranial magnetic stimulation for pain after spinal cord injury: A systematic review and meta-analysis. *J. Neurosurg. Sci.* **2017**, *61*, 514–522. [PubMed]
25. Lefaucheur, J.P.; Ménard-Lefaucheur, I.; Goujon, C.; Keravel, Y.; Nguyen, J.P. Predictive value of rtms in the identification of responders to epidural motor cortex stimulation therapy for pain. *J. Pain* **2011**, *12*, 1102–1111. [CrossRef] [PubMed]
26. André-Obadia, N.; Mertens, P.; Lelekov-Boissard, T.; Afif, A.; Magnin, M.; Garcia-Larrea, L. Is life better after motor cortex stimulation for pain control? Results at long-term and their prediction by preoperative rtms. *Pain Physician* **2014**, *17*, 53–62. [PubMed]
27. Lefaucheur, J.P. Cortical neurostimulation for neuropathic pain: State of the art and perspectives. *Pain* **2016**, *157* (Suppl. 1), S81–S89. [CrossRef]
28. Yoon, E.J.; Kim, Y.K.; Kim, H.R.; Kim, S.E.; Lee, Y.; Shin, H.I. Transcranial direct current stimulation to lessen neuropathic pain after spinal cord injury: A mechanistic pet study. *Neurorehabil. Neural Repair* **2014**, *28*, 250–259. [CrossRef] [PubMed]

29. Guy, S.D.; Mehta, S.; Casalino, A.; Côté, I.; Kras-Dupuis, A.; Moulin, D.E.; Parrent, A.G.; Potter, P.; Short, C.; Teasell, R.; et al. The canpain sci clinical practice guidelines for rehabilitation management of neuropathic pain after spinal cord: Recommendations for treatment. *Spinal Cord* **2016**, *54* (Suppl. 1), S14–S23. [CrossRef] [PubMed]

30. Shealy, C.N.; Mortimer, J.T.; Reswick, J.B. Electrical inhibition of pain by stimulation of the dorsal columns: Preliminary clinical report. *Anesth Analg* **1967**, *46*, 489–491. [CrossRef] [PubMed]

31. Shealy, C.N.; Mortimer, J.T.; Hagfors, N.R. Dorsal column electroanalgesia. *J. Neurosurg.* **1970**, *32*, 560–564. [CrossRef] [PubMed]

32. Melzack, R.; Wall, P.D. Pain mechanisms: A new theory. *Science* **1965**, *150*, 971–979. [CrossRef] [PubMed]

33. Chakravarthy, K.; Kent, A.R.; Raza, A.; Xing, F.; Kinfe, T.M. Burst spinal cord stimulation: Review of preclinical studies and comments on clinical outcomes. *Neuromodulation* **2018**, *21*, 431–439. [CrossRef] [PubMed]

34. Kumar, K.; Toth, C.; Nath, R.K.; Laing, P. Epidural spinal cord stimulation for treatment of chronic pain—Some predictors of success. A 15-year experience. *Surg. Neurol.* **1998**, *50*, 110–120. [CrossRef]

35. Cho, J.H.; Lee, J.H.; Song, K.S.; Hong, J.Y.; Joo, Y.S.; Lee, D.H.; Hwang, C.J.; Lee, C.S. Treatment outcomes for patients with failed back surgery. *Pain Physician* **2017**, *20*, E29–E43. [PubMed]

36. Baber, Z.; Erdek, M.A. Failed back surgery syndrome: Current perspectives. *J. Pain Res.* **2016**, *9*, 979–987. [CrossRef] [PubMed]

37. North, R.B.; Kidd, D.H.; Farrokhi, F.; Piantadosi, S.A. Spinal cord stimulation versus repeated lumbosacral spine surgery for chronic pain: A randomized, controlled trial. *Neurosurgery* **2005**, *56*, 98–106. [CrossRef] [PubMed]

38. Kumar, K.; Taylor, R.S.; Jacques, L.; Eldabe, S.; Meglio, M.; Molet, J.; Thomson, S.; O'Callaghan, J.; Eisenberg, E.; Milbouw, G.; et al. The effects of spinal cord stimulation in neuropathic pain are sustained: A 24-month follow-up of the prospective randomized controlled multicenter trial of the effectiveness of spinal cord stimulation. *Neurosurgery* **2008**, *63*, 762–770. [CrossRef] [PubMed]

39. Kumar, K.; Taylor, R.S.; Jacques, L.; Eldabe, S.; Meglio, M.; Molet, J.; Thomson, S.; O'Callaghan, J.; Eisenberg, E.; Milbouw, G.; et al. Spinal cord stimulation versus conventional medical management for neuropathic pain: A multicentre randomised controlled trial in patients with failed back surgery syndrome. *Pain* **2007**, *132*, 179–188. [CrossRef] [PubMed]

40. Rigoard, P.; Desai, M.J.; North, R.B.; Taylor, R.S.; Annemans, L.; Greening, C.; Tan, Y.; Van den Abeele, C.; Shipley, J.; Kumar, K. Spinal cord stimulation for predominant low back pain in failed back surgery syndrome: Study protocol for an international multicenter randomized controlled trial (promise study). *Trials* **2013**, *14*, 376. [CrossRef] [PubMed]

41. Kemler, M.A.; Barendse, G.A.; van Kleef, M.; de Vet, H.C.; Rijks, C.P.; Furnée, C.A.; van den Wildenberg, F.A. Spinal cord stimulation in patients with chronic reflex sympathetic dystrophy. *N. Engl. J. Med.* **2000**, *343*, 618–624. [CrossRef] [PubMed]

42. Kemler, M.A.; de Vet, H.C.; Barendse, G.A.; van den Wildenberg, F.A.; van Kleef, M. Effect of spinal cord stimulation for chronic complex regional pain syndrome type i: Five-year final follow-up of patients in a randomized controlled trial. *J. Neurosurg.* **2008**, *108*, 292–298. [CrossRef] [PubMed]

43. Slangen, R.; Schaper, N.C.; Faber, C.G.; Joosten, E.A.; Dirksen, C.D.; van Dongen, R.T.; Kessels, A.G.; van Kleef, M. Spinal cord stimulation and pain relief in painful diabetic peripheral neuropathy: A prospective two-center randomized controlled trial. *Diabetes Care* **2014**, *37*, 3016–3024. [CrossRef] [PubMed]

44. De Vos, C.C.; Meier, K.; Zaalberg, P.B.; Nijhuis, H.J.; Duyvendak, W.; Vesper, J.; Enggaard, T.P.; Lenders, M.W. Spinal cord stimulation in patients with painful diabetic neuropathy: A multicentre randomized clinical trial. *Pain* **2014**, *155*, 2426–2431. [CrossRef] [PubMed]

45. Taylor, R.S.; Van Buyten, J.P.; Buchser, E. Spinal cord stimulation for chronic back and leg pain and failed back surgery syndrome: A systematic review and analysis of prognostic factors. *Spine* **2005**, *30*, 152–160. [CrossRef] [PubMed]

46. Tiede, J.; Brown, L.; Gekht, G.; Vallejo, R.; Yearwood, T.; Morgan, D. Novel spinal cord stimulation parameters in patients with predominant back pain. *Neuromodulation* **2013**, *16*, 370–375. [CrossRef] [PubMed]

47. Kapural, L.; Yu, C.; Doust, M.W.; Gliner, B.E.; Vallejo, R.; Sitzman, B.T.; Amirdelfan, K.; Morgan, D.M.; Brown, L.L.; Yearwood, T.L.; et al. Novel 10-khz high-frequency therapy (hf10 therapy) is superior to

traditional low-frequency spinal cord stimulation for the treatment of chronic back and leg pain: The senza-rct randomized controlled trial. *Anesthesiology* **2015**, *123*, 851–860. [CrossRef] [PubMed]

48. Thomson, S.J.; Tavakkolizadeh, M.; Love-Jones, S.; Patel, N.K.; Gu, J.W.; Bains, A.; Doan, Q.; Moffitt, M. Effects of rate on analgesia in kilohertz frequency spinal cord stimulation: Results of the proco randomized controlled trial. *Neuromodulation* **2018**, *21*, 67–76. [CrossRef] [PubMed]

49. Deer, T.; Slavin, K.V.; Amirdelfan, K.; North, R.B.; Burton, A.W.; Yearwood, T.L.; Tavel, E.; Staats, P.; Falowski, S.; Pope, J.; et al. Success using neuromodulation with burst (sunburst) study: Results from a prospective, randomized controlled trial using a novel burst waveform. *Neuromodulation* **2018**, *21*, 56–66. [CrossRef] [PubMed]

50. Kriek, N.; Groeneweg, J.G.; Stronks, D.L.; de Ridder, D.; Huygen, F.J. Preferred frequencies and waveforms for spinal cord stimulation in patients with complex regional pain syndrome: A multicentre, double-blind, randomized and placebo-controlled crossover trial. *Eur. J. Pain* **2017**, *21*, 507–519. [CrossRef] [PubMed]

51. Sdrulla, A.D.; Guan, Y.; Raja, S.N. Spinal cord stimulation: Clinical efficacy and potential mechanisms. *Pain Pract.* **2018**. [CrossRef] [PubMed]

52. Deer, T.R.; Levy, R.M.; Kramer, J.; Poree, L.; Amirdelfan, K.; Grigsby, E.; Staats, P.; Burton, A.W.; Burgher, A.H.; Obray, J.; et al. Dorsal root ganglion stimulation yielded higher treatment success rate for complex regional pain syndrome and causalgia at 3 and 12 months: A randomized comparative trial. *Pain* **2017**, *158*, 669–681. [CrossRef] [PubMed]

53. Mole, J.A.; Prangnell, S.J. Role of clinical neuropsychology in deep brain stimulation: Review of the literature and considerations for clinicians. *Appl. Neuropsychol. Adult* **2017**, 1–14. [CrossRef] [PubMed]

54. Shimizu, T.; Hosomi, K.; Maruo, T.; Goto, Y.; Shimokawa, T.; Haruhiko, K.; Saitoh, Y. Repetitive transcranial magnetic stimulation accuracy as a spinal cord stimulation outcome predictor in patients with neuropathic pain. *J. Clin. Neurosci.* **2018**, *53*, 100–105. [CrossRef] [PubMed]

55. Katayama, Y.; Yamamoto, T.; Kobayashi, K.; Kasai, M.; Oshima, H.; Fukaya, C. Motor cortex stimulation for phantom limb pain: Comprehensive therapy with spinal cord and thalamic stimulation. *Stereotact. Funct. Neurosurg.* **2001**, *77*, 159–162. [CrossRef] [PubMed]

56. Katayama, Y.; Yamamoto, T.; Kobayashi, K.; Kasai, M.; Oshima, H.; Fukaya, C. Motor cortex stimulation for post-stroke pain: Comparison of spinal cord and thalamic stimulation. *Stereotact. Funct. Neurosurg.* **2001**, *77*, 183–186. [CrossRef] [PubMed]

57. Nandi, D.; Smith, H.; Owen, S.; Joint, C.; Stein, J.; Aziz, T. Peri-ventricular grey stimulation versus motor cortex stimulation for post stroke neuropathic pain. *J. Clin. Neurosci.* **2002**, *9*, 557–561. [CrossRef] [PubMed]

58. Neuman, S.A.; Eldrige, J.S.; Hoelzer, B.C. Atypical facial pain treated with upper thoracic dorsal column stimulation. *Clin. J. Pain* **2011**, *27*, 556–558. [CrossRef] [PubMed]

59. Chivukula, S.; Tempel, Z.J.; Weiner, G.M.; Gande, A.V.; Chen, C.J.; Ding, D.; Moossy, J.J. Cervical and cervicomedullary spinal cord stimulation for chronic pain: Efficacy and outcomes. *Clin. Neurol. Neurosurg.* **2014**, *127*, 33–41. [CrossRef] [PubMed]

60. Nandi, D.; Yianni, J.; Humphreys, J.; Wang, S.; O'sullivan, V.; Shepstone, B.; Stein, J.F.; Aziz, T.Z. Phantom limb pain relieved with different modalities of central nervous system stimulation: A clinical and functional imaging case report of two patients. *Neuromodulation* **2004**, *7*, 176–183. [CrossRef] [PubMed]

61. Bittar, R.G.; Yianni, J.; Wang, S.; Liu, X.; Nandi, D.; Joint, C.; Scott, R.; Bain, P.G.; Gregory, R.; Stein, J.; et al. Stereotactic and functional neurosurgery resident award: Deep brain stimulation for generalized dystonia and spasmodic torticollis: Rate and extent of postoperative improvement. *Clin. Neurosurg.* **2005**, *52*, 379–383. [PubMed]

62. Pereira, E.A.; Muthusamy, K.A.; De Pennington, N.; Joint, C.A.; Aziz, T.Z. Deep brain stimulation of the pedunculopontine nucleus in parkinson's disease. Preliminary experience at oxford. *Br. J. Neurosurg.* **2008**, *22* (Suppl. 1), S41–S44. [CrossRef] [PubMed]

63. Boccard, S.G.J.; Prangnell, S.J.; Pycroft, L.; Cheeran, B.; Moir, L.; Pereira, E.A.C.; Fitzgerald, J.J.; Green, A.L.; Aziz, T.Z. Long-term results of deep brain stimulation of the anterior cingulate cortex for neuropathic pain. *World Neurosurg.* **2017**, *106*, 625–637. [CrossRef] [PubMed]

64. Keifer, O.P.; Riley, J.P.; Boulis, N.M. Deep brain stimulation for chronic pain: Intracranial targets, clinical outcomes, and trial design considerations. *Neurosurg. Clin. N. Am.* **2014**, *25*, 671–692. [CrossRef] [PubMed]

65. Hosobuchi, Y.; Rossier, J.; Bloom, F.E.; Guillemin, R. Stimulation of human periaqueductal gray for pain relief increases immunoreactive beta-endorphin in ventricular fluid. *Science* **1979**, *203*, 279–281. [CrossRef] [PubMed]

66. Fessler, R.G.; Brown, F.D.; Rachlin, J.R.; Mullan, S.; Fang, V.S. Elevated beta-endorphin in cerebrospinal fluid after electrical brain stimulation: Artifact of contrast infusion? *Science* **1984**, *224*, 1017–1019. [CrossRef] [PubMed]

67. Sims-Williams, H.; Matthews, J.C.; Talbot, P.S.; Love-Jones, S.; Brooks, J.C.; Patel, N.K.; Pickering, A.E. Deep brain stimulation of the periaqueductal gray releases endogenous opioids in humans. *Neuroimage* **2017**, *146*, 833–842. [CrossRef] [PubMed]

68. Maarrawi, J.; Peyron, R.; Mertens, P.; Costes, N.; Magnin, M.; Sindou, M.; Laurent, B.; Garcia-Larrea, L. Brain opioid receptor density predicts motor cortex stimulation efficacy for chronic pain. *Pain* **2013**, *154*, 2563–2568. [CrossRef] [PubMed]

69. Maarrawi, J.; Peyron, R.; Mertens, P.; Costes, N.; Magnin, M.; Sindou, M.; Laurent, B.; Garcia-Larrea, L. Motor cortex stimulation for pain control induces changes in the endogenous opioid system. *Neurology* **2007**, *69*, 827–834. [CrossRef] [PubMed]

70. Pereira, E.A.; Wang, S.; Peachey, T.; Lu, G.; Shlugman, D.; Stein, J.F.; Aziz, T.Z.; Green, A.L. Elevated gamma band power in humans receiving naloxone suggests dorsal periaqueductal and periventricular gray deep brain stimulation produced analgesia is opioid mediated. *Exp. Neurol.* **2013**, *239*, 248–255. [CrossRef] [PubMed]

71. Mohammadi, A.; Mehdizadeh, A.R. Deep brain stimulation and gene expression alterations in parkinson's disease. *J. Biomed. Phys. Eng.* **2016**, *6*, 47–50. [PubMed]

72. Tilley, D.M.; Cedeño, D.L.; Kelley, C.A.; Benyamin, R.; Vallejo, R. Spinal cord stimulation modulates gene expression in the spinal cord of an animal model of peripheral nerve injury. *Reg. Anesth Pain Med.* **2016**, *41*, 750–756. [CrossRef] [PubMed]

73. Olds, J.; Milner, P. Positive reinforcement produced by electrical stimulation of septal area and other regions of rat brain. *J. Comp. Physiol. Psychol.* **1954**, *47*, 419–427. [CrossRef] [PubMed]

74. Heath, R.G. Psychiatry. *Annu. Rev. Med.* **1954**, *5*, 223–236. [CrossRef] [PubMed]

75. Gol, A. Relief of pain by electrical stimulation of the septal area. *J. Neurol. Sci.* **1967**, *5*, 115–120. [CrossRef]

76. Sweet, W.H.; Wepsic, J.G. Treatment of chronic pain by stimulation of fibers of primary afferent neuron. *Trans. Am. Neurol. Assoc.* **1968**, *93*, 103–107. [PubMed]

77. Reynolds, D.V. Surgery in the rat during electrical analgesia induced by focal brain stimulation. *Science* **1969**, *164*, 444–445. [CrossRef] [PubMed]

78. Mayer, D.J.; Wolfle, T.L.; Akil, H.; Carder, B.; Liebeskind, J.C. Analgesia from electrical stimulation in the brainstem of the rat. *Science* **1971**, *174*, 1351–1354. [CrossRef] [PubMed]

79. Richardson, D.E.; Akil, H. Long term results of periventricular gray self-stimulation. *Neurosurgery* **1977**, *1*, 199–202. [CrossRef] [PubMed]

80. Richardson, D.E.; Akil, H. Pain reduction by electrical brain stimulation in man. Part 1: Acute administration in periaqueductal and periventricular sites. *J. Neurosurg.* **1977**, *47*, 178–183. [CrossRef] [PubMed]

81. Richardson, D.E.; Akil, H. Pain reduction by electrical brain stimulation in man. Part 2: Chronic self-administration in the periventricular gray matter. *J. Neurosurg.* **1977**, *47*, 184–194. [CrossRef] [PubMed]

82. Hosobuchi, Y.; Adams, J.E.; Linchitz, R. Pain relief by electrical stimulation of the central gray matter in humans and its reversal by naloxone. *Science* **1977**, *197*, 183–186. [CrossRef] [PubMed]

83. Mark, V.H.; Ervin, F.R.; Hackett, T.P. Clinical aspects of stereotactic thalamotomy in the human. Part i. The treatment of chronic severe pain. *Arch. Neurol.* **1960**, *3*, 351–367. [CrossRef] [PubMed]

84. Mark, V.H.; Ervin, F.R. Role of thalamotomy in treatment of chronic severe pain. *Postgrad. Med.* **1965**, *37*, 563–571. [CrossRef] [PubMed]

85. Ervin, F.R.; Brown, C.E.; Mark, V.H. Striatal influence on facial pain. *Confin. Neurol.* **1966**, *27*, 75–90. [CrossRef] [PubMed]

86. Mazars, G.; Mérienne, L.; Ciolocca, C. Intermittent analgesic thalamic stimulation. Preliminary note. *Rev. Neurol.* **1973**, *128*, 273–279. [PubMed]

87. Mazars, G.; Roge, R.; Mazars, Y. Results of the stimulation of the spinothalamic fasciculus and their bearing on the physiopathology of pain. *Rev Neurol* **1960**, *103*, 136–138. [PubMed]

88. Mazars, G.; Merienne, L.; Cioloca, C. Treatment of certain types of pain with implantable thalamic stimulators. *Neurochirurgie* **1974**, *20*, 117–124. [PubMed]

89. Mazars, G.J. Intermittent stimulation of nucleus ventralis posterolateralis for intractable pain. *Surg. Neurol.* **1975**, *4*, 93–95. [CrossRef]

90. Hosobuchi, Y.; Adams, J.E.; Rutkin, B. Chronic thalamic stimulation for the control of facial anesthesia dolorosa. *Arch. Neurol.* **1973**, *29*, 158–161. [CrossRef] [PubMed]

91. Hosobuchi, Y.; Adams, J.E.; Rutkin, B. Chronic thalamic and internal capsule stimulation for the control of central pain. *Surg. Neurol.* **1975**, *4*, 91–92. [CrossRef]

92. Adams, J.E.; Hosobuchi, Y.; Fields, H.L. Stimulation of internal capsule for relief of chronic pain. *J. Neurosurg.* **1974**, *41*, 740–744. [CrossRef] [PubMed]

93. Fields, H.L.; Adams, J.E. Pain after cortical injury relieved by electrical stimulation of the internal capsule. *Brain* **1974**, *97*, 169–178. [CrossRef] [PubMed]

94. Thoden, U.; Doerr, M.; Dieckmann, G.; Krainick, J.U. Medial thalamic permanent electrodes for pain control in man: An electrophysiological and clinical study. *Electroencephalogr. Clin. Neurophysiol.* **1979**, *47*, 582–591. [CrossRef]

95. Andy, O.J. Parafascicular-center median nuclei stimulation for intractable pain and dyskinesia (painful-dyskinesia). *Appl. Neurophysiol.* **1980**, *43*, 133–144. [CrossRef] [PubMed]

96. Ray, C.D.; Burton, C.V. Deep brain stimulation for severe, chronic pain. *Acta Neurochir. Suppl.* **1980**, *30*, 289–293. [PubMed]

97. Boivie, J.; Meyerson, B.A. A correlative anatomical and clinical study of pain suppression by deep brain stimulation. *Pain* **1982**, *13*, 113–126. [CrossRef]

98. Levy, R.M.; Lamb, S.; Adams, J.E. Treatment of chronic pain by deep brain stimulation: Long term follow-up and review of the literature. *Neurosurgery* **1987**, *21*, 885–893. [CrossRef] [PubMed]

99. Marchand, S.; Kupers, R.C.; Bushnell, M.C.; Duncan, G.H. Analgesic and placebo effects of thalamic stimulation. *Pain* **2003**, *105*, 481–488. [CrossRef]

100. Coffey, R.J. Deep brain stimulation for chronic pain: Results of two multicenter trials and a structured review. *Pain Med.* **2001**, *2*, 183–192. [CrossRef] [PubMed]

101. Fontaine, D.; Hamani, C.; Lozano, A. Efficacy and safety of motor cortex stimulation for chronic neuropathic pain: Critical review of the literature. *J. Neurosurg.* **2009**, *110*, 251–256. [CrossRef] [PubMed]

102. Boccard, S.G.; Pereira, E.A.; Moir, L.; Aziz, T.Z.; Green, A.L. Long-term outcomes of deep brain stimulation for neuropathic pain. *Neurosurgery* **2013**, *72*, 221–230. [CrossRef] [PubMed]

103. Kumar, K.; Toth, C.; Nath, R.K. Deep brain stimulation for intractable pain: A 15-year experience. *Neurosurgery* **1997**, *40*, 736–746. [CrossRef] [PubMed]

104. Rasche, D.; Rinaldi, P.C.; Young, R.F.; Tronnier, V.M. Deep brain stimulation for the treatment of various chronic pain syndromes. *Neurosurg. Focus* **2006**, *21*, E8. [CrossRef] [PubMed]

105. Son, B.C.; Kim, D.R.; Kim, H.S.; Lee, S.W. Simultaneous trial of deep brain and motor cortex stimulation in chronic intractable neuropathic pain. *Stereotact Funct. Neurosurg.* **2014**, *92*, 218–226. [CrossRef] [PubMed]

106. Marques, A.; Chassin, O.; Morand, D.; Pereira, B.; Debilly, B.; Derost, P.; Ulla, M.; Lemaire, J.J.; Durif, F. Central pain modulation after subthalamic nucleus stimulation: A crossover randomized trial. *Neurology* **2013**, *81*, 633–640. [CrossRef] [PubMed]

107. Lempka, S.F.; Malone, D.A.; Hu, B.; Baker, K.B.; Wyant, A.; Ozinga, J.G.; Plow, E.B.; Pandya, M.; Kubu, C.S.; Ford, P.J.; et al. Randomized clinical trial of deep brain stimulation for poststroke pain. *Ann. Neurol.* **2017**, *81*, 653–663. [CrossRef] [PubMed]

108. Sears, N.C.; Machado, A.G.; Nagel, S.J.; Deogaonkar, M.; Stanton-Hicks, M.; Rezai, A.R.; Henderson, J.M. Long-term outcomes of spinal cord stimulation with paddle leads in the treatment of complex regional pain syndrome and failed back surgery syndrome. *Neuromodulation* **2011**, *14*, 312–318. [CrossRef] [PubMed]

109. Kamano, S. Author's experience of lateral medullary infarction—Thermal perception and muscle allodynia. *Pain* **2003**, *104*, 49–53. [CrossRef]

110. Pereira, E.A.; Lu, G.; Wang, S.; Schweder, P.M.; Hyam, J.A.; Stein, J.F.; Paterson, D.J.; Aziz, T.Z.; Green, A.L. Ventral periaqueductal grey stimulation alters heart rate variability in humans with chronic pain. *Exp. Neurol.* **2010**, *223*, 574–581. [CrossRef] [PubMed]

111. Pereira, E.A.; Wang, S.; Paterson, D.J.; Stein, J.F.; Aziz, T.Z.; Green, A.L. Sustained reduction of hypertension by deep brain stimulation. *J. Clin. Neurosci.* **2010**, *17*, 124–127. [CrossRef] [PubMed]

112. Horn, A.; Neumann, W.J.; Degen, K.; Schneider, G.H.; Kühn, A.A. Toward an electrophysiological "Sweet spot" For deep brain stimulation in the subthalamic nucleus. *Hum. Brain Mapp.* **2017**, *38*, 3377–3390. [CrossRef] [PubMed]

113. Viswanathan, A.; Harsh, V.; Pereira, E.A.; Aziz, T.Z. Cingulotomy for medically refractory cancer pain. *Neurosurg. Focus* **2013**, *35*, E1. [CrossRef] [PubMed]

114. Pereira, E.A.; Paranathala, M.; Hyam, J.A.; Green, A.L.; Aziz, T.Z. Anterior cingulotomy improves malignant mesothelioma pain and dyspnoea. *Br. J. Neurosurg.* **2014**, *28*, 471–474. [CrossRef] [PubMed]

115. Cosgrove, G.R.; Rauch, S.L. Stereotactic cingulotomy. *Neurosurg. Clin. N. Am.* **2003**, *14*, 225–235. [CrossRef]

116. Foltz, E.L.; White, L.E. Pain "Relief" By frontal cingulumotomy. *J. Neurosurg.* **1962**, *19*, 89–100. [CrossRef] [PubMed]

117. Ballantine, H.T.; Cassidy, W.L.; Flanagan, N.B.; Marino, R. Stereotaxic anterior cingulotomy for neuropsychiatric illness and intractable pain. *J. Neurosurg.* **1967**, *26*, 488–495. [CrossRef] [PubMed]

118. Yen, C.P.; Kung, S.S.; Su, Y.F.; Lin, W.C.; Howng, S.L.; Kwan, A.L. Stereotactic bilateral anterior cingulotomy for intractable pain. *J. Clin. Neurosci.* **2005**, *12*, 886–890. [CrossRef] [PubMed]

119. Assaf, Y.; Pasternak, O. Diffusion tensor imaging (dti)-based white matter mapping in brain research: A review. *J. Mol. Neurosci.* **2008**, *34*, 51–61. [CrossRef] [PubMed]

120. Osaka, N.; Osaka, M.; Morishita, M.; Kondo, H.; Fukuyama, H. A word expressing affective pain activates the anterior cingulate cortex in the human brain: An fmri study. *Behav. Brain Res.* **2004**, *153*, 123–127. [CrossRef] [PubMed]

121. Davis, K.D.; Taub, E.; Duffner, F.; Lozano, A.M.; Tasker, R.R.; Houle, S.; Dostrovsky, J.O. Activation of the anterior cingulate cortex by thalamic stimulation in patients with chronic pain: A positron emission tomography study. *J. Neurosurg.* **2000**, *92*, 64–69. [CrossRef] [PubMed]

122. Spooner, J.; Yu, H.; Kao, C.; Sillay, K.; Konrad, P. Neuromodulation of the cingulum for neuropathic pain after spinal cord injury. Case report. *J. Neurosurg.* **2007**, *107*, 169–172. [CrossRef] [PubMed]

123. Boccard, S.G.; Pereira, E.A.; Moir, L.; Van Hartevelt, T.J.; Kringelbach, M.L.; FitzGerald, J.J.; Baker, I.W.; Green, A.L.; Aziz, T.Z. Deep brain stimulation of the anterior cingulate cortex: Targeting the affective component of chronic pain. *Neuroreport* **2014**, *25*, 83–88. [CrossRef] [PubMed]

124. Boccard, S.G.; Fitzgerald, J.J.; Pereira, E.A.; Moir, L.; Van Hartevelt, T.J.; Kringelbach, M.L.; Green, A.L.; Aziz, T.Z. Targeting the affective component of chronic pain: A case series of deep brain stimulation of the anterior cingulate cortex. *Neurosurgery* **2014**, *74*, 628–635. [CrossRef] [PubMed]

125. Maslen, H.; Cheeran, B.; Pugh, J.; Pycroft, L.; Boccard, S.; Prangnell, S.; Green, A.L.; FitzGerald, J.; Savulescu, J.; Aziz, T. Unexpected complications of novel deep brain stimulation treatments: Ethical issues and clinical recommendations. *Neuromodulation* **2018**, *21*, 135–143. [CrossRef] [PubMed]

126. Carroll, D.; Joint, C.; Maartens, N.; Shlugman, D.; Stein, J.; Aziz, T.Z. Motor cortex stimulation for chronic neuropathic pain: A preliminary study of 10 cases. *Pain* **2000**, *84*, 431–437. [CrossRef]

127. Radic, J.A.; Beauprie, I.; Chiasson, P.; Kiss, Z.H.; Brownstone, R.M. Motor cortex stimulation for neuropathic pain: A randomized cross-over trial. *Can. J. Neurol. Sci.* **2015**, *42*, 401–409. [CrossRef] [PubMed]

128. Sachs, A.J.; Babu, H.; Su, Y.F.; Miller, K.J.; Henderson, J.M. Lack of efficacy of motor cortex stimulation for the treatment of neuropathic pain in 14 patients. *Neuromodulation* **2014**, *17*, 303–310. [CrossRef] [PubMed]

129. Lefaucheur, J.P.; Drouot, X.; Cunin, P.; Bruckert, R.; Lepetit, H.; Créange, A.; Wolkenstein, P.; Maison, P.; Keravel, Y.; Nguyen, J.P. Motor cortex stimulation for the treatment of refractory peripheral neuropathic pain. *Brain* **2009**, *132*, 1463–1471. [CrossRef] [PubMed]

130. Nguyen, J.P.; Velasco, F.; Brugières, P.; Velasco, M.; Keravel, Y.; Boleaga, B.; Brito, F.; Lefaucheur, J.P. Treatment of chronic neuropathic pain by motor cortex stimulation: Results of a bicentric controlled crossover trial. *Brain Stimul.* **2008**, *1*, 89–96. [CrossRef] [PubMed]

131. Velasco, F.; Argüelles, C.; Carrillo-Ruiz, J.D.; Castro, G.; Velasco, A.L.; Jiménez, F.; Velasco, M. Efficacy of motor cortex stimulation in the treatment of neuropathic pain: A randomized double-blind trial. *J. Neurosurg.* **2008**, *108*, 698–706. [CrossRef] [PubMed]

132. Velasco, F.; Carrillo-Ruiz, J.D.; Castro, G.; Argüelles, C.; Velasco, A.L.; Kassian, A.; Guevara, U. Motor cortex electrical stimulation applied to patients with complex regional pain syndrome. *Pain* **2009**, *147*, 91–98. [CrossRef] [PubMed]

133. Rasche, D.; Ruppolt, M.; Stippich, C.; Unterberg, A.; Tronnier, V.M. Motor cortex stimulation for long-term relief of chronic neuropathic pain: A 10 year experience. *Pain* **2006**, *121*, 43–52. [CrossRef] [PubMed]

134. Monsalve, G.A. Motor cortex stimulation for facial chronic neuropathic pain: A review of the literature. *Surg. Neurol. Int.* **2012**, *3*, S290–S311. [CrossRef] [PubMed]

135. Rasche, D.; Tronnier, V.M. Clinical significance of invasive motor cortex stimulation for trigeminal facial neuropathic pain syndromes. *Neurosurgery* **2016**, *79*, 655–666. [CrossRef] [PubMed]

136. Honey, C.M.; Tronnier, V.M.; Honey, C.R. Deep brain stimulation versus motor cortex stimulation for neuropathic pain: A minireview of the literature and proposal for future research. *Comput. Struct. Biotechnol. J.* **2016**, *14*, 234–237. [CrossRef] [PubMed]

137. Lilford, R.J.; Thornton, J.G.; Braunholtz, D. Clinical trials and rare diseases: A way out of a conundrum. *BMJ* **1995**, *311*, 1621–1625. [CrossRef] [PubMed]

138. Owen, S.L.; Green, A.L.; Nandi, D.; Bittar, R.G.; Wang, S.; Aziz, T.Z. Deep brain stimulation for neuropathic pain. *Neuromodulation* **2006**, *9*, 100–106. [CrossRef] [PubMed]

139. Owen, S.L.; Green, A.L.; Stein, J.F.; Aziz, T.Z. Deep brain stimulation for the alleviation of post-stroke neuropathic pain. *Pain* **2006**, *120*, 202–206. [CrossRef] [PubMed]

140. Pereira, E.A.; Boccard, S.G.; Aziz, T.Z. Deep brain stimulation for pain: Distinguishing dorsolateral somesthetic and ventromedial affective targets. *Neurosurgery* **2014**, *61* (Suppl. 1), 175–181. [CrossRef] [PubMed]

141. Green, A.L.; Wang, S.; Stein, J.F.; Pereira, E.A.; Kringelbach, M.L.; Liu, X.; Brittain, J.S.; Aziz, T.Z. Neural signatures in patients with neuropathic pain. *Neurology* **2009**, *72*, 569–571. [CrossRef] [PubMed]

142. Kringelbach, M.L.; Jenkinson, N.; Green, A.L.; Owen, S.L.; Hansen, P.C.; Cornelissen, P.L.; Holliday, I.E.; Stein, J.; Aziz, T.Z. Deep brain stimulation for chronic pain investigated with magnetoencephalography. *Neuroreport* **2007**, *18*, 223–228. [CrossRef] [PubMed]

143. Ray, N.J.; Jenkinson, N.; Kringelbach, M.L.; Hansen, P.C.; Pereira, E.A.; Brittain, J.S.; Holland, P.; Holliday, I.E.; Owen, S.; Stein, J.; et al. Abnormal thalamocortical dynamics may be altered by deep brain stimulation: Using magnetoencephalography to study phantom limb pain. *J. Clin. Neurosci.* **2009**, *16*, 32–36. [CrossRef] [PubMed]

144. Pereira, E.A.; Green, A.L.; Bradley, K.M.; Soper, N.; Moir, L.; Stein, J.F.; Aziz, T.Z. Regional cerebral perfusion differences between periventricular grey, thalamic and dual target deep brain stimulation for chronic neuropathic pain. *Stereotact. Funct. Neurosurg.* **2007**, *85*, 175–183. [CrossRef] [PubMed]

145. Hentall, I.D.; Luca, C.C.; Widerstrom-Noga, E.; Vitores, A.; Fisher, L.D.; Martinez-Arizala, A.; Jagid, J.R. The midbrain central gray best suppresses chronic pain with electrical stimulation at very low pulse rates in two human cases. *Brain Res.* **2016**, *1632*, 119–126. [CrossRef] [PubMed]

Review

Spinal Cord Stimulation for Neuropathic Pain: Current Trends and Future Applications

Ivano Dones * and Vincenzo Levi

Neurosurgery Department, Functional Neurosurgery, Unit for the Surgical Treatment of Pain and Spasticity, Fondazione Istituto Neurologico Carlo Besta, 20131 Milan, Italy; vincenzo.levi@unimi.it
* Correspondence: ivanodones@gmail.com; Tel.: +39-34931-81075

Received: 30 June 2018; Accepted: 18 July 2018; Published: 24 July 2018

Abstract: The origin and the neural pathways involved in chronic neuropathic pain are still not extensively understood. For this reason, despite the wide variety of pain medications available on the market, neuropathic pain is challenging to treat. The present therapeutic alternative considered as the gold standard for many kinds of chronic neuropathic pain is epidural spinal cord stimulation (SCS). Despite its proved efficacy, the favourable cost-effectiveness when compared to the long-term use of poorly effective drugs and the expanding array of indications and technical improvements, SCS is still worldwide largely neglected by general practitioners, neurologists, neurosurgeons and pain therapists, often bringing to a large delay in considering as a therapeutic option for patients affected by neuropathic chronic pain. The present state of the art of SCS in the treatment of chronic neuropathic pain is here overviewed and speculations on whether to use a trial period or direct implant, to choose between percutaneous leads or paddle electrodes and on the pros and cons of the different patterns of stimulation presently available on the market (tonic stim, high-frequency stim and burst stim) are described.

Keywords: spinal cord stimulation (SCS); neuromodulation; neuropathic pain

---

1. Introduction

The International Association for the Study of Pain defines neuropathic pain as the pain caused by a lesion or disorder of the somatosensory nervous system. It affects 7–10% of the general population, entailing an overall physical and psychological burden more relevant than that seen with nociceptive pain [1]. Although given the continuous development of new molecules appearing on the market to control neuropathic pain, this invalidating symptom is currently poorly improved by available drug treatments. In addition, common analgesic and opioid therapies carry a non-negligible risk of adverse events in the long term. Alternative options were lesional surgery at the dorsal root entry zone and, more recently, a number of neuromodulation procedures [2]. Among them, spinal cord stimulation (SCS) or dorsal column stimulation constitutes an advanced neuromodulation procedure able to actually decrease neuropathic pain in many syndromes such as in failed back surgery syndrome (FBSS), complex regional pain syndrome (CRPS) type I and II, postherpetic neuralgia and pure radicular pain [3]. Despite its proved efficacy, the favourable cost-effectiveness, when compared to the long-term use of poorly effective drugs and the expanding array of indications and technical improvements, spinal cord stimulation (SCS) is still worldwide largely neglected by general practitioners, neurologists, neurosurgeons and pain therapists, often bringing to a large delay in considering as a therapeutic option for patients affected by neuropathic chronic pain [4,5].

The present state of the art of SCS in the treatment of chronic neuropathic pain is overviewed and speculations on whether to use a trial period or direct implant, to choose between percutaneous leads or paddle electrodes and on the pros and cons of the different patterns of stimulation presently available on the market (tonic stim, high-frequency stim and burst stim) are described.

2. Mechanism of Action

SCS owes its inception to the gate control theory (GCT), theorized by Wall and Melzack in their seminal 1965 paper. Wall and Melzack speculated that the nociceptive signal would be inhibited by antidromic activation of collateral, large, myelinated Aß fibres in the dorsal columns [6]. The first reported clinical application of dorsal column stimulation came 2 years later and at the time SCS was thought to act merely at the spinal segmental level [7]. However, The GCT theory did not take into account two evident SCS contradictions. The first is that, accordingly to the theory, SCS should be more effective in controlling acute nociceptive pain, which in fact is not the case. Secondly, Wall and Melzack's theory is not able to explain the pain-free interval that is often noticed after discontinuation of stimulation [8]. For these reasons, the GCT theory seems to get more and more inconsistent to explain the mechanism of action of spinal cord stimulation in favour of other hypotheses, some of which involve the supraspinal pathway of pain control and transmission [9]. A pain-modulating dorsal column–brainstem–spinal loop was recently identified in animal models, while neuroimaging studies demonstrate that tonic SCS mainly acts by modulating the lateral pain ascending pathway and by interfering with the electrical and metabolic activity of the cingulate gyrus, lateral sensory thalamic nuclei, prefrontal cortex and postcentral gyrus [10,11]. Sato and colleagues showed that analgesic properties of SCS could be hampered by the use of opioid antagonists, thus suggesting that SCS might be also effective through the activation of the descending opioid pathway [12].

Several other experimental studies have elucidated the role of different transmitter systems which would be enhanced or inhibited by tonic dorsal column stimulation. SCS is deemed to neutralize the overexcitability of wide dynamic range (WDR) neurons in the dorsal horn by increasing γ-amino-butyric acid (GABA) release [13]. WDR neuron wind-up caused by excessive nociceptive inputs is believed to trigger the lateral pain pathway, giving the start to the abnormal transmission of pain sensation to the brain. So far, it remains unclear whether the SCS rebalancing effect of the system occurs solely as a result of presynaptic inhibition of the WDR neurons via antidromic activation or if it is due to most complex combined pre/postsynaptic phenomena [14]. Finally, evidence in experimental models in rats also suggests a role of the cholinergic transmitter systems. Increase in acetylcholine release was noticed under SCS even in association to the activation of the M4 muscarinic receptors, while low doses of muscarinic receptor agonist led to enhance the SCS-induced analgesic effect in rats [15,16].

However, the exact mechanism that allows the improvement of neuropathic pain observed in a large percentage of patients submitted to SCS still remains so far unclear.

3. SCS Indications and Patient Selection

Over the last years, a growing number of chronic pain syndromes of neuropathic origin have been treated with SCS, from brachial plexus and peripheral nerve injuries to postherpetic neuralgia and central pain of spinal cord origin, with varying grade of evidences [17] (Table 1).

Table 1. Common SCS indications and contraindications.

SCS Common Neuropathic Indications	SCS Main Contraindications
Failed back surgery syndrome	
Complex regional pain syndrome (I and II)	Infection
Radicular and nerve root pain	Coagulopathy
Postherpetic neuralgia	Spinal stenosis
Pain due to peripheral nerve injury	Psychiatric disorders
Intercostal neuralgia	Substance abuse
Phantom pain	

SCS: spinal cord stimulation.

To date, however, there are only two clinical pain syndromes that clearly benefit from SCS treatment: the failed back surgery syndrome (FBSS) and type 1 and 2 of the complex regional pain syndrome (CRPS). In Europe as well as in the United States, FBSS represents the most common indication for an SCS implant. Patients affected by FBSS are those who did not achieve satisfying outcome after single or multiple spinal operations in terms of pain relief, or who developed new, recurrent, drug-resistant low back or radicular pain regardless of the surgical procedure and possible surgical malpractice. This condition is often underrated, thus largely procrastinating the possible implant of SCS in favour of repetitive surgical procedures on the patient's spine. A rate of recurrent back or leg pain of 5–36% in patients who had lumbar disc herniation surgery at a 2-year follow-up was recently reported in literature, whereas a prospective study by Skolasky et al. involving 260 patients who underwent surgical laminectomy with or without fusion for lumbar spinal stenosis secondary to degenerative alterations showed that 29.2% of patients had either no change or, even, increased pain at the 12-month follow-up after surgery [18,19]. Although there are only a few high-quality, large prospective and randomized comparative trials reported, the literature on SCS reports a large number of case series but only evidence supporting the use of SCS for the treatment of FBSS are of real significance. A systematic and comprehensive review regarding the effectiveness of SCS in treating chronic spinal pain showed that there is a clear (Level I–II) role for conventional low-frequency SCS as a treatment for otherwise intractable lumbar FBSS [20]. In another recent and extensive meta-analysis about conventional SCS for chronic back and leg pain, more than half of the patients experienced remarkable pain relief, independently from previous spinal surgery the patients possibly underwent. The pain remission was maintained during a mean follow-up period of 24 months [21]. In an exhaustive and thorough literature review, Cameron found an overall success rate of 62% among the 747 patients affected by FBSS and treated with SCS [22].

Although of minor incidence, particularly if compared to FBSS, the treatment of CRPS by SCS is also well established and includes one randomized controlled trial (RCT), which compared in a cohort of 54 patients SCS plus physical therapy with physical therapy alone [23]. At 6 months, in the SCS group pain was reduced by 3.6 on the visual analogue scale (VAS), while in the group receiving physical therapy alone, VAS was increased by 0.2 ($p < 0.001$). No clinically relevant improvement in functional status after 6-months follow-up was detected. The health-related quality of life improved only in 24 patients who underwent implantation of a spinal cord stimulator. Recently, the randomized prospective ACCURATE trial has compared SCS with another promising neuromodulation technique, the Dorsal Root Ganglion Stimulation (DRGS), which involves the percutaneous placement of a lead in the epidural posterosuperior space of the intervertebral foramen [24]. Both methods have been proved effective, but a higher statistical significance was associated with DRGS when considering pain relief, postural stability and mood improvement. Though efficacy of DRGS for CRPS treatment seems favorable, this surgical option is still in its inception. In addition, the data from the ACCURATE trial still need to be replicated, whereas there is more high-quality evidence to support the use of SCS [25].

Given the multitude of growing indications, appropriate patient selection is of paramount importance to achieve best SCS efficacy.

Patients who underwent surgical spine procedures may still suffer from unrecognized persistent compression of the neural elements. For this reason, a pre-operative spinal magnetic resonance imaging (MRI) should routinely be performed to search for an organic substrate of the pain. In that case, the patient should be considered for reoperation, otherwise SCS may be proposed as the next therapeutic option. Distinguishing neuropathic pain from other causes of pain may also be challenging. Over recent years, several helpful screening tools for the correct diagnosis of neuropathic pain have been validated. Among them, the Neuropathic Pain Questionnaire (NPQ), ID Pain and PainDETECT are widely available and easily deliverable, relying on interview questions only [26–29]. Finally, many guidelines have claimed the importance of a pre-operative psychological evaluation. This step may be precious for two reasons. Firstly, it greatly helps in excluding patients in whom a coexistence of major psychiatric diseases such as major depression, psychosis or drug abuse may hamper their response to stimulation. It must be said, however, that whether some studies found a negative association between depression and response to SCS, others did not, the evidence in literature thus being quite discordant [30–32].

In the case of appropriate indication and experienced implanter, SCS success rates are generally remarkable (in the range of about 50–75%). Despite all the careful selection pearls mentioned above, however, a variable percentage of patients do not benefit from SCS independently from the appropriateness of the indication. The cause of that partial response is still unclear; as also unclear is the average decrease in pain of 50% in the responders. Therefore, in order to increase the success rate of the procedure, a two-step surgery consisting of a trial stimulation phase before definitive internal pulse generator (IPG) implantation has become standard practice in most centres since the first SCS introduction. Kumar et al. found that about 17–20% of the patients decide not to proceed with the implantation although the trial period induced complete paraesthesia coverage of the painful region [33]. If the trial stimulation period has the indisputable advantage of avoiding probable unsuccessful implantation, on the other hand it carries a non-negligible risk of infection, which is reported to be between 2.4% and 18.6%, with consequent need of hardware removal and antibiotic therapy [34,35]. This wide range of infection rate reported is explained taking into account different factors, such as the single-centre surgical volume and surgeon experience [36]. Besides the risk of infection, the real utility and predictive value of the trial phase compared to direct permanent implantation has never been established through prospective, randomized, controlled trials. To date, the only paper addressing the topic is an Italian multicentre study enrolling 122 patients. In this paper, the authors assessed long-term clinical SCS efficacy in patients who were submitted to a trial period and in patients who, on the contrary, underwent immediate permanent implant. Significant reduction in pain, as measured by variation in visual analogue scale (VAS) score, was observed at least 1 year after implantation in both groups. Surprisingly, SCS efficacy was greater in patients who underwent permanent implant at once (59.5% vs. 71.4%). This difference, however, was not statistically relevant [37].

4. Technical Nuances

Despite the overall mini-invasive nature and straightforwardness of the procedure in experienced hands, several points regarding SCS surgical technique are still not standardized worldwide and deserve further discussion. One of these pertains to the choice between general and local anaesthesia. Many centres worldwide usually prefer to have the patient under local anaesthesia, using percutaneous-type electrode with the aid of fluoroscopic guidance. This strategy carries some clear advantages, the more obvious of which is to prevent any possible complication due to general anaesthesia and open surgery. In addition, many SCS experts claim that a percutaneous surgery in the awake patient is recommendable as it gives the opportunity to test the patient response to stimulation through his direct confirmation of full-paraesthesia coverage, thus confirming the correct electrode positioning [17]. Awake lead placement, however, has some shortcomings. It is well known that the success of any kind of awake surgery largely depends on the patient's collaboration. Several individual patient's factors, such as anxiety, stress or discomfort, should be taken into account and carefully assessed during the pre-operative screening before proceeding to awake surgery. Hence, a not-negligible portion of patients might not be able to tolerate the procedure. Moreover, awake surgery usually allows percutaneous leads to be easily positioned, whereas paddle lead requires a more invasive laminectomy approach in many cases. On the contrary, a lead electrode can move and get dislodged with body movements. In addition, it is well known from the literature that although the lead electrode is properly positioned, there may be a slight-to-moderate loss of SCS efficacy over time (in the range of 25–50%) due to both minor dislodgements of the electrode and the formation of scar tissue all around the leads [31,32]. Under these circumstances, a paddle lead might be more useful than a percutaneous one being steadily in contact with the dural surface and with negligible tilting or dislodgement in the long term. Moreover, a lead electrode produces a spherical electric field of which only the part toward the dural surface is effective. Conversely, a paddle electrode's electric field is oriented to the spinal cord only, thus it needs less electric power to obtain similar results. Furthermore, when there is a need to have a widespread laterally extended electric field, one single paddle electrode with multiple lines of contacts can be used to shape the proper electric field that, on the contrary, with a lead electrode could be obtained only by positioning two separate electrodes. The Neurostimulation Appropriateness Consensus Committee (NACC) guidelines recently stated that "Confirmation of correct lead placement has been advocated with either awake intraoperative confirmation of paraesthesia coverage or use of neuromonitoring in asleep placement, such as Electromyography (EMG) responses or Somatosensory evoked potential (SSEP) collision testing." [38]. To date, there is only a single prospective, multicentre study comparing safety and efficacy of the neuromonitoring-assisted asleep SCS implantation technique as compared to conscious procedures [39]. The authors found that SCS placement under general anaesthesia was a shorter procedure with superior paraesthesia coverage profiles, while maintaining lower adverse events and equal clinical outcomes for pain relief compared to awake surgery.

SCS is usually regarded as a safe procedure due to its reversible and minimally invasive characteristics [40]. Severe adverse events, such as spinal epidural bleeding and permanent neurologic deficit, are rare, whereas hardware complication and infection has been reported with an incidence of 24–50% and 7.5%, respectively [41–43] (Table 2).

Table 2. SCS surgical complications.

SCS Common Complications
More frequent
Hardware-related (lead migration, breakage, connection failure, malfunctioning, pain at the IPG)
Haematoma and seroma at IPG site
Rare
Spinal epidural haematoma
CSF leak
Neurological deficit

IPG: internal pulse generator; CSF: cerebrospinal fluid.

It has been recently pointed out that the type of lead used may have an impact on both hardware complication and infection rate. In a single-centre prospective nonrandomized trial, Kinfe et al. compared effectiveness and safety of both lead types in a cohort of 100 patients who underwent SCS for FBSS with a 2-year follow-up [44]. They found a comparable clinical efficacy, but higher dislocation and infection rates in the group with cylindrical electrodes (14% and 10%, respectively) than in the group with paddle electrodes (6% and 2%, respectively). Another paper comprehensively analysed a large, independent, cohort of 131,774 patients from the United States who underwent percutaneous or paddle lead SCS placement comparing the incidence of complications, reoperation rates, and medication health-care costs both for percutaneous and paddle lead [45]. Placement of paddle leads was associated to a slightly higher initial postoperative complications, but with a significantly lower long-term reoperation rates. On the contrary, no difference in terms of health-care costs was noticed. Finally, in a retrospective study in a large cohort of 8326 patients conducted by Petraglia and colleagues [46], no significant difference in the rates of spinal cord trauma or spinal hematoma was observed between the two types of lead. In conclusion, current available data indicate an overall comparable and acceptable clinical efficacy and safety for both percutaneous and paddle lead. At the moment, the choice between the two types seems to mainly rely on individual implanter preference and background, while the choice between temporary implant and definite implant seems irrelevant in terms of percentage of good results in the long term, provided that an accurate selection of patients has been done.

5. Current and Future Development

The last decade saw an exponential technological advancement in the whole field of neuromodulation. Particularly, SCS therapy took advantage of the introduction of rechargeable generators, multiple leads paddle electrodes, position-sensing stimulation and MRI compatible devices that represent well-established great innovations [5]. Recently, this continuous innovative trend brought to an ongoing revolution on new different patterns of electric stimulation. Conventional SCS is based on a tonic pulse, released at constant frequency, (40–80 Hz), and a fixed pulse width of 200–450 μs and varying current amplitudes tailored on any patient needs [47,48]. This modality of stimulation is effective, but with a variable percentage of success that hardly is higher than 50% of pain control (in some series, a 50% pain relief is reported in approximately 50% of patients) and with a frequent progressive tolerance in the long term [21]. Consequently, there is an increasing need for new stimulation patterns, aimed both at improving SCS results in non-responders and avoiding long-term adaptation to the electrical therapy. In this respect, several types of new electric parameters are currently extensively investigated. Burst stimulation and high-frequency stimulation are the two main new stimulation options available so far. De Ridder et al. published a cohort of 12 patients who underwent the so-called "burst stimulation" [49]. This new stimulation pattern consists of intermittent trains of five high-frequency stimuli delivered at 500 Hz, 40 times per second and with a long pulse width and an interspike interval of 1000 μs delivered in constant-current mode. The monophasic pulses are charge-balanced at the end of the burst, differentiating it from clustered high-frequency tonic firing [50]. Applying this stimulation pattern, De Ridder and colleagues found that, when compared to conventional SCS, burst stimulation gave remarkable long-term pain higher suppression with a concomitant greater reduction in the number of patients sensing paraesthesias due to stimulation (92% vs. 17% of patient, respectively). In addition, a major extension of the stimulation effect to the midline region seems, unlike conventional tonic stimulation, to be observed during burst stimulation. This major advantage seems to be ascribed to the higher chance of intercepting even deep nerve fibres by means of trains of impulses at higher frequency [51].

Another paraesthesia-free technique is the high-frequency continuous stimulation. High-frequency stimulation is similar in principle to tonic stimulation, using 30 μs pulse width and individually actively charge-balanced pulses delivered at very high frequency (10 kHz) [47,52]. It is based on the staggered implantation of two 8-contact electrodes at the thoracic level (T8 down to T12) and although the reason for its inception is still unclear, it is thought to decrease WDR neuron firing rates and their consequent wind-up phenomenon [53]. At the moment, however, this assumption is not supported by any experimental or clinical evidence [8]. Regarding its clinical efficacy, in a recent prospective multicentre study 70% of patients treated by high-frequency SCS experienced a significant and sustained low back pain and leg pain relief, greater than 50%, without referring any concomitant induced paraesthesia [53]. On the other hand, no significant differences were found between a short trial period (2 weeks) of sham stimulation and high-frequency 5 kHz stimulation in a randomized study including 33 patients [54].

Given these considerations, it is not yet possible to draw final conclusions on the real pain-relieving efficiency of both burst and high-frequency stimulation. Although these new stimulation modalities seem very promising, further prospective randomized clinical trials are needed to prove their presumed clinical superiority over conventional tonic SCS.

6. Conclusions

Although still underused, conventional SCS may be considered as an effective, safe, well-tolerated and reversible treatment option for severe drug-refractory neuropathic pain. Accurate indications and cautious patient selection represent the principal mainstays for the success of this treatment. In the near future, there will surely be confirmations as to the efficacy of the new patterns of stimulation both at high frequency and through burst stimulation and, possibly, future new patterns to improve the efficacy of this treatment in improving chronic neuropathic pain.

Author Contributions: I.D. and V.L. contributed equally to the conception and design of the article, analysis and interpretation. Both authors were involved in critically drafting/revising the article for important intellectual content.

Funding: This research received no external funding.

Conflicts of Interest: All the authors declare that they have no financial or other conflicts of interest in relation to this research and its publication.

References

1. Torrance, N.; Smith, B.H.; Bennett, M.I.; Lee, A.J. The epidemiology of chronic pain of predominantly neuropathic origin. Results from a general population survey. *J. Pain* **2006**, *7*, 281–289. [CrossRef] [PubMed]
2. Finnerup, N.B.; Otto, M.; McQuay, H.J.; Jensen, T.S.; Sindrup, S.H. Algorithm for neuropathic pain treatment: an evidence based proposal. *Pain* **2005**, *118*, 289–305. [CrossRef] [PubMed]
3. Lee, A.W.; Pilitsis, J.G. Spinal cord stimulation: Indications and outcomes. *Neurosurg. Focus* **2006**, *21*, 1–6. [CrossRef]
4. Hoelscher, C.; Riley, J.; Wu, C.; Sharan, A. Cost-Effectiveness Data Regarding Spinal Cord Stimulation for Low Back Pain. *Spine* **2017**, *42* (Suppl. 14), S72–S79. [CrossRef] [PubMed]
5. Slavin, K.V. Spinal Stimulation for Pain: Future Applications. *Neurotherapeutics* **2014**, *11*, 535–542. [CrossRef] [PubMed]
6. Melzack, R.; Wall, P.D. Pain mechanisms: A new theory. *Science* **1965**, *150*, 971–979. [CrossRef] [PubMed]
7. Shealy, C.N.; Mortimer, J.T.; Reswick, J.B. Electrical inhibition of pain by stimulation of the dorsal columns: Preliminary clinical report. *Anesth. Analg.* **1967**, *46*, 489–491. [CrossRef] [PubMed]
8. Wolter, T. Spinal cord stimulation for neuropathic pain: Current perspectives. *J. Pain Res.* **2014**, *7*, 651–663. [CrossRef] [PubMed]
9. Barchini, J.; Tchachaghian, S.; Shamaa, F.; Jabbur, S.J.; Meyerson, B.A.; Song, Z.; Linderoth, B.; Saade, N.E. Spinal segmental and supraspinal mechanisms underlying the pain-relieving effects of spinal cord stimulation: An experimental study in a rat model of neuropathy. *Neuroscience* **2012**, *215*, 196–208. [CrossRef] [PubMed]
10. El-Khoury, C.; Hawwa, N.; Baliki, M.; Atweh, S.F.; Jabbur, S.J.; Saade, N.E. Attenuation of neuropathic pain by segmental and supraspinal activation of the dorsal column system in awake rats. *Neuroscience* **2002**, *112*, 541–553. [CrossRef]
11. Rasche, D.; Siebert, S.; Stippich, C.; Kress, B.; Nennig, E.; Sartor, K.; Tronnier, V.M. Spinal cord stimulation in Failed-Back-Surgery-Syndrome. Preliminary study for the evaluation of therapy by functional magnetic resonance imaging (fMRI). *Schmerz* **2005**, *19*, 497–500, 502–505. [CrossRef] [PubMed]
12. Sato, K.L.; King, E.W.; Johanek, L.M.; Sluka, K.A. Spinal cord stimulation reduces hypersensitivity through activation of opioid receptors in a frequency-dependent manner. *Eur. J. Pain* **2013**, *17*, 551–561. [CrossRef] [PubMed]
13. Yakhnitsa, V.; Linderoth, B.; Meyerson, B.A. Spinal cord stimulation attenuates dorsal horn neuronal hyperexcitability in a rat model of mononeuropathy. *Pain* **1999**, *79*, 223–233. [CrossRef]
14. Foreman, R.D.; Linderoth, B. Neural mechanisms of spinal cord stimulation. *Int. Rev. Neurobiol.* **2012**, *107*, 87–119. [CrossRef] [PubMed]
15. Schechtmann, G.; Song, Z.; Ultenius, C.; Meyerson, B.A.; Linderoth, B. Cholinergic mechanisms involved in the pain relieving effect of spinal cord stimulation in a model of neuropathy. *Pain* **2008**, *139*, 136–145. [CrossRef] [PubMed]
16. Song, Z.; Meyerson, B.A.; Linderoth, B. Muscarinic receptor activation potentiates the effect of spinal cord stimulation on pain-related behavior in rats with mononeuropathy. *Neurosci. Lett.* **2008**, *436*, 7–12. [CrossRef] [PubMed]
17. Nagel, S.J.; Lempka, S.F.; Machado, A.G. Percutaneous spinal cord stimulation for chronic pain: Indications and patient selection. *Neurosurg. Clin. N. Am.* **2014**, *25*, 723–733. [CrossRef] [PubMed]
18. Parker, S.L.; Mendenhall, S.K.; Godil, S.S.; Sivasubramanian, P.; Cahill, K.; Ziewacz, J.; McGirt, M.J. Incidence of Low Back Pain After Lumbar Discectomy for Herniated Disc and Its Effect on Patient-reported Outcomes. *Clin. Orthop. Relat. Res.* **2015**, *473*, 1988–1999. [CrossRef] [PubMed]

19. Skolasky, R.L.; Wegener, S.T.; Maggard, A.M.; Riley, L.H. The impact of reduction of pain after lumbar spine surgery: The relationship between changes in pain and physical function and disability. *Spine* **2014**, *39*, 1426–1432. [CrossRef] [PubMed]

20. Grider, J.S.; Manchikanti, L.; Carayannopoulos, A.; Sharma, M.L.; Balog, C.C.; Harned, M.E.; Grami, V.; Justiz, R.; Nouri, K.H.; Hayek, S.M.; et al. Effectiveness of Spinal Cord Stimulation in Chronic Spinal Pain: A Systematic Review. *Pain Physician* **2016**, *19*, E33–E54. [PubMed]

21. Taylor, R.S.; Desai, M.J.; Rigoard, P.; Taylor, R.J. Predictors of pain relief following spinal cord stimulation in chronic back and leg pain and failed back surgery syndrome: A systematic review and meta-regression analysis. *Pain Pract.* **2014**, *14*, 489–505. [CrossRef] [PubMed]

22. Cameron, T. Safety and efficacy of spinal cord stimulation for the treatment of chronic pain: A 20-year literature review. *J. Neurosurg.* **2004**, *100*, 254–267. [CrossRef] [PubMed]

23. Kemler, M.A.; Barendse, G.A.; Van Kleef, M.; De Vet, H.C.; Rijks, C.P.; Furnee, C.A.; Van den Wildenberg, F.A. Spinal cord stimulation in patients with chronic reflex sympathetic dystrophy. *N. Engl. J. Med.* **2000**, *343*, 618–624. [CrossRef] [PubMed]

24. Deer, T.R.; Levy, R.M.; Kramer, J.; Poree, L.; Amirdelfan, K.; Grigsby, E.; Staats, P.; Burton, A.W.; Burgher, A.H.; Obray, J.; et al. Dorsal root ganglion stimulation yielded higher treatment success rate for complex regional pain syndrome and causalgia at 3 and 12 months: A randomized comparative trial. *Pain* **2017**, *158*, 669–681. [CrossRef] [PubMed]

25. Harrison, C.; Epton, S.; Bojanic, S.; Green, A.L.; FitzGerald, J.J. The Efficacy and Safety of Dorsal Root Ganglion Stimulation as a Treatment for Neuropathic Pain: A Literature Review. *Neuromodulation* **2017**, 225–233. [CrossRef] [PubMed]

26. Bennett, M.I.; Attal, N.; Backonja, M.M.; Baron, R.; Bouhassira, D.; Freynhagen, R.; Scholz, J.; Tolle, T.R.; Wittchen, H.U.; Jensen, T.S. Using screening tools to identify neuropathic pain. *Pain* **2007**, *127*, 199–203. [CrossRef] [PubMed]

27. Krause, S.J.; Backonja, M.M. Development of a neuropathic pain questionnaire. *Clin. J. Pain* **2003**, *19*, 306–314. [CrossRef] [PubMed]

28. Portenoy, R. Development and testing of a neuropathic pain screening questionnaire: ID Pain. *Curr. Med. Res. Opin.* **2006**, *22*, 1555–1565. [CrossRef] [PubMed]

29. Freynhagen, R.; Baron, R.; Gockel, U.; Tolle, T.R. painDETECT: A new screening questionnaire to identify neuropathic components in patients with back pain. *Curr. Med. Res. Opin.* **2006**, *22*, 1911–1920. [CrossRef] [PubMed]

30. Burchiel, K.J.; Anderson, V.C.; Brown, F.D.; Fessler, R.G.; Friedman, W.A.; Pelofsky, S.; Weiner, R.L.; Oakley, J.; Shatin, D. Prospective, multicenter study of spinal cord stimulation for relief of chronic back and extremity pain. *Spine* **1996**, *21*, 2786–2794. [CrossRef] [PubMed]

31. Wolter, T.; Fauler, I.; Kieselbach, K. The impact of psychological factors on outcomes for spinal cord stimulation: An analysis with long-term follow-up. *Pain Physician* **2013**, *16*, 265–275. [PubMed]

32. Sparkes, E.; Raphael, J.H.; Duarte, R.V.; LeMarchand, K.; Jackson, C.; Ashford, R.L. A systematic literature review of psychological characteristics as determinants of outcome for spinal cord stimulation therapy. *Pain* **2010**, *150*, 284–289. [CrossRef] [PubMed]

33. Kumar, K.; Wilson, J.R. Factors affecting spinal cord stimulation outcome in chronic benign pain with suggestions to improve success rate. *Acta Neurochir. Suppl.* **2007**, *97*, 91–99. [PubMed]

34. May, M.S.; Banks, C.; Thomson, S.J. A Retrospective, Long-term, Third-Party Follow-up of Patients Considered for Spinal Cord Stimulation. *Neuromodulation* **2002**, *5*, 137–144. [CrossRef] [PubMed]

35. Hoelzer, B.C.; Bendel, M.A.; Deer, T.R.; Eldrige, J.S.; Walega, D.R.; Wang, Z.; Costandi, S.; Azer, G.; Qu, W.; Falowski, S.M.; et al. Spinal Cord Stimulator Implant Infection Rates and Risk Factors: A Multicenter Retrospective Study. *Neuromodulation* **2017**, *20*, 558–562. [CrossRef] [PubMed]

36. Rudiger, J.; Thomson, S. Infection rate of spinal cord stimulators after a screening trial period. A 53-month third party follow-up. *Neuromodulation* **2011**, *14*, 136–141. [CrossRef] [PubMed]

37. Colombo, E.V.; Mandelli, C.; Mortini, P.; Messina, G.; De Marco, N.; Donati, R.; Irace, C.; Landi, A.; Lavano, A.; Mearini, M.; et al. Epidural spinal cord stimulation for neuropathic pain: A neurosurgical multicentric Italian data collection and analysis. *Acta Neurochir.* **2015**, *157*, 711–720. [CrossRef] [PubMed]

38. Deer, T.R.; Lamer, T.J.; Pope, J.E.; Falowski, S.M.; Provenzano, D.A.; Slavin, K.; Golovac, S.; Arle, J.; Rosenow, J.M.; Williams, K.; et al. The Neurostimulation Appropriateness Consensus Committee (NACC) Safety Guidelines for the Reduction of Severe Neurological Injury. *Neuromodulation* **2017**, *20*, 15–30. [CrossRef] [PubMed]

39. Falowski, S.M.; Sharan, A.; McInerney, J.; Jacobs, D.; Venkatesan, L.; Agnesi, F. Nonawake vs. Awake Placement of Spinal Cord Stimulators: A Prospective, Multicenter Study Comparing Safety and Efficacy. *Neurosurgery* **2018**, 1–8. [CrossRef] [PubMed]

40. Bendersky, D.; Yampolsky, C. Is spinal cord stimulation safe? A review of its complications. *World Neurosurg.* **2014**, *82*, 1359–1368. [CrossRef] [PubMed]

41. Levy, R.M. Device complication and failure management in neuromodulation. *Neuromodulation* **2013**, *16*, 495–502. [CrossRef] [PubMed]

42. Eldabe, S.; Buchser, E.; Duarte, R.V. Complications of Spinal Cord Stimulation and Peripheral Nerve Stimulation Techniques: A Review of the Literature. *Pain Med.* **2016**, *17*, 325–336. [CrossRef] [PubMed]

43. Shamji, M.F.; Westwick, H.J.; Heary, R.F. Complications related to the use of spinal cord stimulation for managing persistent postoperative neuropathic pain after lumbar spinal surgery. *Neurosurg. Focus* **2015**, *39*, E15. [CrossRef] [PubMed]

44. Kinfe, T.M.; Quack, F.; Wille, C.; Schu, S.; Vesper, J. Paddle versus cylindrical leads for percutaneous implantation in spinal cord stimulation for failed back surgery syndrome: A single-center trial. *J. Neurol. Surg. A Cent. Eur. Neurosurg.* **2014**, *75*, 467–473. [CrossRef] [PubMed]

45. Babu, R.; Hazzard, M.A.; Huang, K.T.; Ugiliweneza, B.; Patil, C.G.; Boakye, M.; Lad, S.P. Outcomes of percutaneous and paddle lead implantation for spinal cord stimulation: A comparative analysis of complications, reoperation rates, and health-care costs. *Neuromodulation* **2013**, *16*, 418–426. [CrossRef] [PubMed]

46. Petraglia, F.W.; Farber, S.H.; Gramer, R.; Verla, T.; Wang, F.; Thomas, S.; Parente, B.; Lad, S.P. The incidence of spinal cord injury in implantation of percutaneous and paddle electrodes for spinal cord stimulation. *Neuromodulation* **2016**, *19*, 85–89. [CrossRef] [PubMed]

47. Shechter, R.; Yang, F.; Xu, Q.; Cheong, Y.K.; He, S.Q.; Sdrulla, A.; Carteret, A.F.; Wacnik, P.W.; Dong, X.; Meyer, R.A.; et al. Conventional and kilohertz-frequency spinal cord stimulation produces intensity and frequency-dependent inhibition of mechanical hypersensitivity in a rat model of neuropathic pain. *Anesthesiology* **2013**, *119*, 422–432. [CrossRef] [PubMed]

48. Yearwood, T.L.; Hershey, B.; Bradley, K.; Lee, D. Pulse width programming in spinal cord stimulation: A clinical study. *Pain Physician* **2010**, *13*, 321–335. [PubMed]

49. De Ridder, D.; Vanneste, S.; Plazier, M.; Van der Loo, E.; Menovsky, T. Burst spinal cord stimulation: Toward paresthesia-free pain suppression. *Neurosurgery* **2010**, *66*, 986–990. [CrossRef] [PubMed]

50. Van Havenbergh, T.; Vancamp, T.; Van Looy, P.; Vanneste, S.; De Ridder, D. Spinal cord stimulation for the treatment of chronic back pain patients: 500-hz vs. 1000-hz burst stimulation. *Neuromodulation* **2015**, *18*, 9–12. [CrossRef] [PubMed]

51. Chakravarthy, K.; Kent, A.R.; Raza, A.; Xing, F.; Kinfe, T.M. Burst Spinal Cord Stimulation: Review of Preclinical Studies and Comments on Clinical Outcomes. *Neuromodulation* **2018**. [CrossRef] [PubMed]

52. De Ridder, D.; Perera, S.; Vanneste, S. Are 10 kHz Stimulation and Burst Stimulation Fundamentally the Same? *Neuromodulation* **2017**, *20*, 650–653. [CrossRef] [PubMed]

53. Van Buyten, J.P.; Al-Kaisy, A.; Smet, I.; Palmisani, S.; Smith, T. High-frequency spinal cord stimulation for the treatment of chronic back pain patients: results of a prospective multicenter European clinical study. *Neuromodulation* **2013**, *16*, 56–59. [CrossRef] [PubMed]

54. Perruchoud, C.; Eldabe, S.; Batterham, A.M.; Madzinga, G.; Brookes, M.; Durrer, A.; Rosato, M.; Bovet, N.; West, S.; Bovy, M.; et al. Analgesic efficacy of high-frequency spinal cord stimulation: A randomized double-blind placebo-controlled study. *Neuromodulation* **2013**, *16*, 363–369. [CrossRef] [PubMed]

MDPI

St. Alban-Anlage 66

4052 Basel

Switzerland

Tel. +41 61 683 77 34

Fax +41 61 302 89 18

www.mdpi.com

Brain Sciences Editorial Office

E-mail: brainsci@mdpi.com

www.mdpi.com/journal/brainsci

9 783039 369508